The 80/20 Principle and 92 Other Powerful Laws of Nature

To Lee

Praise

The 80/20 Principle a...
Other Powerful Laws of Nature

"This book boils down the insights from science, from Newton to now, to compile a practical manual for success in business and life. It's a terrific *tour-de-force*."
James A Lawrence, Executive Vice-President, General Mills

"Intriguing and very unusual. Koch is saying things that really need saying, but that you don't hear said elsewhere. Both intellectual and practical, the book is above all highly readable and thought provoking."
Dr Jules Goddard, Fellow, London Business School

"Exciting stuff! *The Power Laws* touches on many of the most important debates going on in academic and business circles, and yet goes beyond them in generating ideas for anyone in business. A major shift in the way we think about the corporation and management is imminent, and this book makes a major contribution to fresh thinking."
Dr Marcus Alexander, Director, Ashridge Strategic Management Centre

"The theory of business genes alone is a good reason for buying this book. And we get sixteen other major scientific principles thrown in as well! The content is academically respectable and often state-of-the-art, but what is particularly impressive is how accessible it all is. Clear and simple, but also authoritative and bursting with insight."
Dr Peter Johnson, Fellow, Exeter College, Oxford

"I never knew that science could be so useful in business! A real revelation."

Ch  ltants

Evolution by Natural Selection—Business Genes—Gause's Laws
Evolutionary Psychology—the Prisoner's Dilemma—Newton's Laws
Relativity—Quantum Mechanics—Chaos—Complexity
The 80/20 Principle—Punctuated Equilibrium—The Tipping Point
Increasing Returns—The Paradox of Enrichment—Entropy
Unintended Consequences

Fisher's fundamental theorem of natural selection—the experience curve—time-based competition—Ulam's dilemma—Mendel's laws of hereditary—DNA and its structure—memes—lifelines—the Hardy-Weinberg law—ecological niches and MacArthur's warblers—owners and intruders—the endowment effect—the ultimatum bargaining game—cultural evolution in animals—neuroplasticity—game theory—the red queen effect—the evolutionary arms race—Ridley's theory of social coagulation division of labor—Ricardo's theory of comparative advantage—theory of co-opetition Linus' law—Diamond's principle of intermediate fragmentation—black holes—Gödel's incompleteness theorem—the medium is the message—Heisenberg's uncertainty principle—the principle of complementarity—Schrödinger's cat—sensitive dependence on initial conditions—the butterfly effect—fractal similarities—the principle of impotence—the hinge factor—first mover advantage—emergence—self-organizing systems—the edge of chaos—Zipf's rank/size law—Simon's theory of clumps and lumps—Gutenberg-Richer law—Parkinson's laws—Cyert & March's theory of organizational slack—landscapes—Zipf's principle of least effort—Juran's rule of the vital few—Von Foerster's theorem—the 50/5 principle—Mendeleev's periodic table control theory—Fermat's principle of least time—trichotomy law—plague theory crossing the chasm—exponential growth—Fibonacci's rabbits—big bang—Say's law of economic arbitrage—Freud's reality principle—law of diminishing returns—Moore's law—Metcalfe's law—the theory of industry sweet spots—Murphy's laws—theory of the second best—system dynamics

The 80/20 Principle and 92 Other Powerful Laws of Nature

Richard Koch

NICHOLAS BREALEY
PUBLISHING
London • Boston

This new edition first published by
Nicholas Brealey Publishing in 2014

3-5 Spafield Street
Clerkenwell, London
EC1R 4QB, UK
Tel: +44 (0)20 7239 0360
Fax: +44 (0)20 7239 0370

20 Park Plaza
Boston
MA 02116, USA
Tel: (888) BREALEY
Fax: (617) 523 3708

www.nicholasbrealey.com

First published as *The Power Laws* (UK) and
The Power Laws of Business (US) in 2000

ISBN 978-1-85788-611-5
eISBN 978-1-85788-918-5

British Library Cataloguing in Publication Data
A catalogue record for this book is available from the British Library.

Printed in Finland by Bookwell.

Contents

The Power Laws

Each of the 12 chapters of this book covers a major power law or a cluster of major power laws, of which there are 17.[1] Each chapter contains a clear message that works brilliantly well in business, and often in life generally. If you turn back to the Contents page, you can see the major power law(s) described for each chapter, followed by the main action implication. When you read the chapters themselves, you will also find other, less major, power laws described, together with some other action implications (and, of course, a detailed explanation of the major action implication). There is a total of 93 power laws, which are listed below.

In the text, **_Major Power Laws_** are in bold italics with the initial letter of each word capitalized, and **_Minor power laws_** are also in bold italics, but with only the first word capitalized.

In order of appearance

Note

1   In my tally of 17 major power laws or clusters thereof, I have counted
Gause's three laws as one cluster, Newton's three laws of motion and his law
of gravity as one cluster, and Einstein's special and general laws of relativity
as one cluster. The total tally of 93 power laws, on the other hand, counts
each of these laws separately (e.g., Gause has three power laws).

Preface to the New Edition

I am sometimes asked to write a new book on the future of business. But I always reply the same way: I have already written the book – it is the one you hold in your hands.

The late Peter Drucker always used to say that the future was already here, its seeds evident in the present. I think about this another way: I see a new paradigm of business struggling to emerge from the husk of the old. Using the word 'paradigm' risks seeming overblown, but there is no other word to describe the magnitude of the change, because it is both theoretical in the best sense and also intensely practical.

The raw material for the new paradigm lies in the body of this book – the most powerful ideas derived from science. There are no fewer than 92 scientific laws or principles that fit all of the following criteria:

◆ They are acute observations, hypotheses, or 'laws' that frequently or always apply.
◆ They are not obvious, and are often counterintuitive.
◆ They are of great value in business, careers, and/or our personal lives.
◆ They are not as well known as they should be, and not used anywhere near as often as they should be.
◆ They have hidden depth and are worth exploring beyond their surface value.

The key elements of the new approach are clearly visible, although they are at odds with prevailing practice:

◆ Work is about insight, not hours put in. Wealth creation is 1 percent perspiration and 99 percent inspiration. Intuition is more important than rational analysis.

◆ Business is driven by ideas and information. The best ideas create the most money. The best ideas are infinitely recyclable.

◆ What is key is not management or finance or clever strategies – it is the product. Nothing is more important than improving or reimagining the product, and developing a unique new business system to support it.

◆ Less is more. Businesses make breakthroughs when they subtract enough, when everything is taken away except the core. In theory this is easy, but in practice it is rarely done. When it is done the result is a landmark product – the printing press, bicycle, Model T Ford, aspirin, Coca-Cola, television, McDonald's, Disneyland, microchip, personal computer, internet, smartphone, iPod – that shapes the future.

◆ Innovation and natural selection create the vast majority of new value in any economy.

◆ Walled gardens create fortunes for a decade or so; open networks create fortunes forever.

◆ Only exponential growth creates interesting new markets and fortunes. Exponential growth usually comes from simplifying a product, market, or business system that has great sophistication but too little simplicity.

The overall answer is in the final chapter, which draws all the ideas together. If you are the sort of impatient person (like me) who wants the answer quickly, you might want to read that chapter before you start the rest.

Make the most of the feast of ideas in front of you – create something new, unique, and indelibly yours from the luxuriant diversity of our intellectual heritage. You will never again think that your life and work are unimportant.

Richard Koch
Gibraltar, January 2014

Overture

On Appreciating a Wonky World

You don't see something until you have the right metaphor to let you perceive it.

Thomas Kuhn

In search of a few universal principles

The prize-winning biologist and author Edward O Wilson defines science as 'the organized, systematic enterprise that gathers knowledge about the world and condenses the knowledge into testable laws and principles.'

Science presents us with a few universal patterns of how things *really* work, rules and relationships that contain tremendous insight, not just within specific scientific disciplines, but also outside them, in business and life generally. There is a surprisingly small number of these simple, recurrent explanations for complex and apparently unrelated phenomena. 'Nature,' it has been well said, 'is prodigal with details but parsimonious with principles.' I have tried to identify the most important and relevant of these patterns, rules and relationships, which I have called 'power laws.' I should make it clear that I am using power law in a colloquial sense and not in the technical, mathematical sense, where a power law is a quantitative relationship expressed in an equation.

My power laws have to justify three criteria for inclusion here:

◆ The power law must be a coherent theory of how things work, with wide acceptance among scientists.
◆ The power law must transcend the discipline where it originated, and be used in more than one scientific discipline.
◆ The power law must be capable of application to business.

I have ranged far and wide in the hunt for useful power laws. Part One includes not just Darwin's theory, modern genetics, and neo-Darwinist theories, but also power laws from evolutionary psychology, conventional psychology, archaeology, palaeontology, anthropology, ecology, neuroplasticity, and game theory. Part Two draws on Newtonian physics, and the mechanical view of science, on Einstein's theories of relativity, on quantum mechanics, and on mathematics, logic, and philosophy. Part Three examines systems theory, chaos, complexity, and economics, appropriating en route some ideas from cybernetics, probability theory, geology, sociology, epidemiology, history, and from some of the disciplines (especially mathematics) used earlier. Although I have not ignored the humanities, there is a strong bias towards the natural and physical sciences.

I have avoided well-trawled areas, such as the insights into managing people available from industrial psychology, and sought fresh perspectives not currently available in management literature. I have generally avoided management concepts, even where these claim (usually spuriously) some scientific validity. I have also tried—I am sure not wholly successfully—to avoid facile comparisons. There is nothing worse than half-baked business conclusions drawn from quarter-understood scientific ideas.

One example of what I mean is the abuse by certain management writers of quantum mechanics, the twentieth-century revolution in physics, which is probably the greatest triumph of science in that century. Revelations about how submicroscopic, inanimate particles behave have been made the foundation for all kinds of theories about how organizations and society should be organized, in many cases with absolutely no justification in quantum mechanics. Some of these theories are sensible, some wacky, but the link with science is tenuous at best (see Chapter 8).

Another example of woolly thinking is the common parallel drawn between evolution by natural selection and the nature of competition between established corporations. This parallel is simply wrong, betraying a profound lack of understanding of both what Darwin said and how modern business competition works. And yet if Darwin is read carefully and sympathetically, there are some wonderful new insights into how business works (see Chapters 1, 2, and the Finale).

My researcher, Andrej Machacek, and I looked at over 1000 scientific theories and principles which, on first inspection, appeared possibly relevant, before winnowing the list down to 93 power laws. We have allowed into our list not only theories well supported by data, but also empirically observed facts, and a few nonverifiable but resonant concepts. We have included a handful of ideas that offer insight without having any scientific validity, such as Murphy's and Parkinson's laws. The vast majority of the power laws, however, are scientifically respectable, and where they are controversial, attention is drawn to this.

Insight from science for business success

Science is an attempt to explain the world around us. Business is part of this world. Physicists know that the universe is unitary: the same laws apply everywhere, all the time. Scientists working in different disciplines have been helped by theories developed elsewhere. What works in biology also works in economics, in physics, and in psychology, and the other way round. Interdisciplinary sciences such as chaos and complexity observe the same phenomena and the same patterns, equally relevant to meteorology, or financial markets, or geology, or physics, or chemistry, or many other disciplines, and usually capable of similar mathematical expression across all these areas.

The reason that insights and theories from one science work in another is that the universe is more fundamental than our scientific taxonomy. In trying to understand and study things we break them down, but all we are doing is glimpsing, from different angles, the same truly universal principles. My quest has been for power laws that transcend scientific boundaries and defy artificial barriers erected between 'science' and 'business.'

So in trying to gain insight from science, I have first tried to under-stand the science properly, in its own terms, before applying it to busi-ness. This is exactly what a chemist does, for example, in seeking to understand and apply an idea from physics.

The progress from order to chaos

Nineteenth-century science was solid and dependable. Twentieth-century science was surreal, often incomprehensible, and pretty incredible. At the start of the twenty-first century, most of us feel more at home with the scientific world view of the late nineteenth century, and we run our lives accordingly. The nineteenth-century view of science was the culmination of three centuries of progressively increasing degrees of understanding and confidence: educated people felt that they understood how the world worked, and that there would soon be few limits to humanity's dominion over nature. A whole new scientific civilization beckoned.

The twentieth century bit back. As scientists learned more, the uni-verse seemed less predictable, less ordered, more mysterious, and more frightening. Defense mechanisms set in. Brilliant scientists like Albert Einstein reached for their blinkers. The universe just couldn't be that ran-dom, that pointless, that out of sync with reason. And so began an intel-lectual reaction that is still with us. Most of our mental models are still those of nineteenth-century science. It's more comfortable that way. Arguably, we're also more productive with those models.

Let's see how the world became so much easier to understand from the sixteenth to the nineteenth centuries, and how much more difficult thereafter.

In praise of the 'incomparable Mr. Newton'

Perhaps the most important science book ever—Sir Isaac Newton's *Principia Mathematica*—was published in 1687. Newton drew together knowledge that had been simmering for centuries and was then coming to the boil: from the ancient Greeks, from Roger Bacon (a late thirteenth-century Oxford scholar), from Leonardo da Vinci, Galileo, Kepler, from French philosopher René Descartes, and from many other sources. Newton was both the father of modern empirical science and the codifier of the most powerful intellectual framework the world has ever seen—the idea of the clockwork universe.

The Newtonian world was simple and easily understood. Everything could be related to everything else: on earth and in the heavens. Reality comprised machines and parts of machines, all behaving in accordance with a few basic, universal, reliable laws. Science as a total system made sense. God was relegated to the role of wise clockmaker, the chap who wound up the clock—the universe—and then left it to operate on its own according to certain standard operating procedures.

You can see the clockwork universe manifested in the work of Adam Smith and all the classical economists; in Thomas Robert Malthus's thoughts on population and sustainability; in the ideas of the French enlightenment on the 'perfectibility of man', an idea encapsulated by British historian Edward Gibbon, who wrote in 1776 that 'we cannot be certain to what heights the human species may aspire'; in Charles Darwin's theory of evolution by natural selection; in Sigmund Freud's mechanical model of the mind and consciousness; and in all political and social writers from Thomas Hobbes to Karl Marx, John Stuart Mill, Auguste Comte, Vilfredo Pareto and Max Weber. Although many of these thinkers added a teleological or dialectical perspective—which is at odds with a simple, static, clockwork world—their views are all mechanistic and rational. Everything is a machine, everything obeys simple laws, everything fits together, everything can be understood, and everything can be analyzed and reduced to its basic elements. Everything works and has a purpose. People can aspire to control the world, society, and their own nature, because everything is mechanical and intelligence can control mechanical things.

The Newtonian ideology gave people such confidence that they could understand and control the world that this is what they did. The explosion of science, industry, technology and wealth that followed in the next 300 years, which was well beyond any historical precedent and which took us to the brink of material utopia, would have been impossible without faith in the clockwork universe.

It is therefore difficult to overstate Newton's impact on business. One route of influence is directly through engineering and machinery, and the productivity revolutions from 1750 to 2000. Another is through the influence of mechanical models on economics and the way that 'organizations'—a modern word, but a very Newtonian concept—are structured. A third is through the power of analysis: the contribution of numbers, accounting systems, calculators and computers, all of which depend on Newtonian methods.

Nearly all executives and management writers inhabit Newton's world, and with good reason. Yet science has moved on.

Weird and wonderful twentieth-century science

Danish physicist Niels Bohr (1885–1962) was one of the great brains of the twentieth century and perhaps the most important developer of quantum physics, which must rank as one of the most sublime, and counter-intuitive, scientific theories of all time. Bohr used to tell a story about a Jewish theological student who attended three lectures by a famous rabbi. The first lecture was splendid, and the student understood it all. The second lecture was even better; the rabbi clearly understood every word, but it was so subtle and deep that the student couldn't follow it all. The third lecture, though, was the crowning achievement; it was so brilliant that even the rabbi didn't understand it. Bohr said that quantum theory was 'weird': it made him feel like the rabbi at the third lecture.

Quantum theory is so subversive that even Albert Einstein kept trying to prove it wrong. As we'll see in Chapter 8, the micro-world of atoms 'chooses' which state to leap into entirely at random; the precise positions or velocities of electrons cannot be measured; light is both like a wave and like a particle—nothing is real, nothing is predictable, everything is uncertain, and everything is related, mysteriously, to everything else.

The first two decades of the twentieth century also gave us Einstein's theories of relativity, which are extraordinarily difficult to understand; Einstein himself said that only 12 people in the world would understand his general theory. As a result of relativity, we know that space is curved: gravitation is the warping of space and time by physical mass. Time and space are not two dimensions, but one linked frame of reference: time is part of the physical universe.

The scientific theme that there is no objective reality was reinforced in 1931 by the brilliant Austrian scientist Kurt Gödel, who was eccentric to the point of madness. Nonetheless, his incompleteness theorem proved beyond doubt that, even within a simple, formal system like mathematics, you could write down statements that could never be either proved or disproved within the terms of that system. Reality is an invention, not a given. So farewell absolute truth!

Systems thinking, and the developments in chaos and complexity in the last third of the twentieth century took us even further back to the future. It turns out that most things in the world, and certainly some of the most important—including the weather, the brain, cities, economies, history, and people—are 'non-linear systems,' which means that they don't behave in the straightforward way assumed by all scientists from Isaac Newton to the end of the nineteenth century.

Non-linear systems don't have simple causes and effects; they don't behave like mechanical objects; everything is interrelated; equilibrium is elusive and fleeting; small and even trivial causes can have massive effects; control is impossible; prediction is hazardous; simple systems can demonstrate incredibly complex behavior; and complex systems can give rise to very simple behavior. In this weird and wonky world, intelligence, common sense and good intentions are no guarantee of good results; instead, unwelcome and unintended consequences are endemic.

This is a topsy-turvy world where classical, 'Newtonian,' cause-and-effect logic can get you into a lot of trouble. Yet scientists have discovered remarkably consistent laws and patterns that can describe 'chaotic' behavior. Although they take some teasing out, beauty, method and order do exist within the apparent madness and disorder.

A new *gestalt* for business?

Fear not, help is at hand. If we understand the world revealed by modern science, we can stop being slaves to defunct physicists, philosophers and economists, and to dysfunctional genes.

We will understand, for instance, why individuals are badly pro-grammed to work effectively in large organizations; and why—for good and ill—organizations have a will of their own.

A slight but crucial change in perspective will demonstrate that the fundamental unit of value in business is economic information; that the market in economic information is highly imperfect, allowing us to appropriate huge value; that technology drives growth; and that entrepreneurs rather than scientists are the main drivers of technology.

We'll see that a struggle for existence is at the heart of business, but that the struggle is primarily between ideas, not between corporations; that corporate competition is marginal to our economies and to our per-sonal success; and that business is not at all like war.

The power laws tell us that innovation is mandatory, but also predictable, following a seamless process of variation, frequent failure, infrequent success, and further variation—a process eerily reminiscent of natural selection; that most experiments have to fail, and yet that experimentation is essential; and that business is not generally structured for experimentation, wrongly pre-ferring the architecture of the cathedral to that of the bazaar.

The new *gestalt* holds that growth is not difficult to find, but is extremely difficult to perpetuate in one vehicle; that less is more; that influence is generally superior to control; and that we are moving into an era where return on management effort (ROME) is more important than return on capital employed (ROCE), and where corporate ownership has more downside than upside.

The new science explains that most of business is non-linear and unpredictable, yet that different branches of business each follow dis-cernible and distinctive patterns; that there are always a few powerful forces that we can use to our advantage or that will upset our plans; and that success usually emerges when we are looking the other way, but that unexpected successes, if we deign to notice them, can then be deliber-ately nurtured into explosive bonanzas.

We'll see that business often obtains diminishing returns from extra effort and investment, while the most important economic phenomenon at the start of the twenty-first century is increasing returns, where additional investment and command of intellectual property throw off exponentially increasing cash.

We'll learn that either/or thinking is a trap, that tradeoffs can be elided, and that a both/and attitude is the handmaiden of creativity; that there is an infinite number of ways to fail, but there are always also multiple routes to success; and that the opposite of a great business truth is … another great business truth.

Finally, the power laws reveal that business is a book of bets, that only skilful gamblers can consistently win; yet that business is also a series of related transactions, linked together by cooperation, loyalty, networks, serial reciprocity and reputation; and that the richest results, and the satisfaction of our own selfish ends, require us to forgo our own short-term self-interest in order to cooperate with the best cooperators. It is not the meek who shall inherit the earth, nor the aggressive, but rather the cooperative.

These are not random opinions or tentative interpretations of science, nor are they wild extrapolations from it. They are well-grounded inferences from scientific theory, and from the observation of business within the rather novel framework that the power laws donate to it. This framework is superior because it fits both dominant scientific insights and business reality, and because it prescribes a set of actions that work, that lead to success. A final key advantage of the new framework is that it can also accommodate the traditional 'mechanical' view of science and business which, after all, has proved its worth.

The old régime has its place

It is important to take a balanced view of the changes in science in the twentieth century and of the appropriate response of business to these changes.

If by some impossible trick we had only twentieth-century science, and nothing from the Newtonian heritage, we would all be incomparably poorer in the depth and power of our thinking and in our wealth. Newton's science would have been enough to send men to the moon and back and for most practical purposes the inaccuracies in his physics can be safely ignored. It is true that tiny, inanimate particles don't behave at all in a Newtonian way, but this doesn't stop us building bridges as we did in the days before quantum theory. Logic may tell us that truth is always elusive and subjective, but we don't and shouldn't behave in our daily lives as though there were no difference between truth and lies. A world whose science was confined to relativity, quantum theory, modern genetics, systems theory, and chaos and complexity theory would be a strange, inhospitable place. Earth might resemble Douglas Adams' appalling planet where they put all the highly paid people, like management consultants and spin doctors and politicians, who couldn't actually do anything.

We need the 'old stuff' in science. We need engineers and chemists and old-style physicists and doctors. We need mechanistic thinking, analysis, and faith in reason.

And we need these things in business too. We need our balance sheets and budgets, our old-fashioned management by objectives, our planning and monitoring, and our faith—illusory or otherwise—in our ability to control our own fate.

The new scientific view has the merit of greater accuracy and understanding of how the universe works. If it is a less appealing view, on the face of it, that is no reason to behave like ostriches. But there is a downside to understanding: it can paralyze; it can make us give up before we start. The great thing about Newtonian science was that it was activist and optimistic: it drove, and drives, huge numbers of ordinary people to achieve extraordinary results. Control was the watchword: the universe could be understood, and it could be controlled.

We now know that control is *not* possible; the universe has a mind of its own, and will defeat our attempts to order and subdue it. And yet it's still important to try! Fatalism or excessive *laissez-faire* will not lead to what we want. A sophisticated anti-Newtonian philosophy is of much less use than a primitive Newtonian one.

Let me illustrate this by jumping ahead to one of the concepts to emerge from complexity theory, *self-organization*. The theory reveals a stunning and irrefutable tendency of complex systems, like cities or economies or human bodies, to organize themselves from a number of simpler parts and earlier stages. They do this according to certain typical patterns, that are repeated, with minor variations, over and over again.

It is undeniable that a business organization is a similar sort of entity: it is a self-organizing system. The simplistic, modernist prescription might therefore be that we should leave organizations to organize themselves. And there is much in this. Anyone who has tried to organize a team from a preordained plan with prescribed roles for each team member knows the limitations of this approach. It's far better to tell the team what to do and let the team members work out their own roles and how to do it.

Yet the extrapolation of this liberal approach to a whole organization— on the implicit grounds that if this is how nature arranges things, this is how we should do it too—is deeply flawed. If it is left alone, the organization will organize itself effectively—for its own ends. It won't do what the its owners or leaders want it to do. Nor will it be functional from society's viewpoint. The self-organizing organization will end up larger and fatter than it needs to be to achieve any given economic objective. This criticism, it is true, comes from an old-fashioned, Newtonian, mechanistic view of the world: it is part of an ideology of control and rational objectives. But if I am accused of harboring this ideology, I gladly plead guilty. The ideology of control and objectives is one price of progress.

Escaping obsolete mental models

Scientists working with relativity, or quantum theory, or modern mathematics, or systems theory, or chaos, or complexity are at the top of their fields. They may not reach absolute truth, but they are closer to knowing what happens and, to a large extent, how and why. But what about the rest of us, trying to pilot our way through life in general and our business affairs in particular? We're sitting ducks. We're bound to misunderstand what is happening, to see most of our efforts lead nowhere much, to pull levers that don't work, and to do things that may lead precisely to the outcomes we most want to avoid. We work in the twenty-first century world using nineteenth-century mental models and, probably, governed by genes that have not changed essentially since the Stone Age.[1]

Yet there is a way out. If we understand a handful of power laws, *and if we act to exploit those laws*, we can multiply our effectiveness.

The power laws of the universe are like the winds. If we're sailing, we have to use the winds, because there is no other source of power on a yacht. But a good sailor doesn't allow the winds to blow her off course. Even against a head wind, she makes progress. She has a map. She has an objective, which is different from that of the winds. She tacks and turns, following a zigzag course that, however tortuous and slow, will bring her safely to port.

We have no other sources of power than those provided by the universe, our own brains and instincts included. We need to understand the power laws, whether these control tiny particles, huge planets, or our own behavior. But we don't then simply say, 'Great, groovy baby!' We respect the laws. We recognize when they can undo our plans. We harness their power in creative ways. But we don't slavishly obey the laws or worship them. We have our own inner light, guiding our faltering steps even as we understand how difficult it is to overcome our programming.

We need a good dose of Newtonian mechanics, Cartesian faith in reason, Gibbonian faith in the perfectibility of man, Darwinian faith in evolution, Marxian faith in our ability to arrange society, and Freudian belief in our ability to control our emotions—all faiths that are intellectually untenable, at least in their extreme forms—while simultaneously understanding and using the weirder and subtler reaches of more recent knowledge.

On with the show!

Note

1   The new science of evolutionary psychology suggests that we're still 'hardwired' for life on the Savannah Plain, and that our emotional responses, though well suited to life 200,000 years ago, are quite at odds with what is needed today. Yet there is also evidence that we can tamper with our own hardwiring; see Chapter 4.

Part One

The Biological Laws

How Economic Information Drives Progress

Introduction to Part One

Part One relates to insights from biology and related disciplines: how life originates, how it is structured, and how it develops and adapts to the conditions around it. Its focus is on the evolution of life, with particular attention to human life, and the relationship between human evolution and business.

Chapter 1 examines Darwin's theory of *Evolution by Natural Selection*, which we have come to take for granted but which is the most amazing, awesome, and counterintuitive way that could be imagined for generating life of ever greater beauty and complexity.

Chapter 2 constructs the *Theory of Business Genes*, where economic information evolves by selection and where replicators—the business genes—seek vehicles for their survival and proliferation.

Chapter 3 looks at ecological niches and the experiments on small organisms by Soviet scientist G F Gause in the 1930s. *Gause's Laws* reinforce the importance of differentiation for business genes and their vehicles.

Chapter 4 covers *Evolutionary Psychology* and the mismatch between our primitive genes and the requirements of modern business. *Punctuated Equilibrium*, a key power law discussed at length in Chapter 11, is also introduced here.

Chapter 5 develops a theory of human cooperation and competition, based on insights from the *Prisoner's Dilemma*, other concepts from game theory, and from biology, economics and anthropology. Selfish objectives, it transpires, can only be met by ever greater degrees of cooperation and interdependence.

1

On Evolution by Natural Selection

If I could give an award for the best idea ever I would give it to Darwin, because his idea unites in a stroke these two completely disparate worlds, until then, of the meaningless mechanical physical sciences, astronomy, physics and chemistry on the one side, and the world of meaning, culture, art and biology.

Daniel Dennett

The universe is run by selection

In the material world, nothing is more important than **Evolution by Natural Selection**. Without natural selection, our species could not exist. If selection did not apply to ideas, technologies, markets, companies, teams and products in precisely the same way as it applies to species, we would all be working on the land struggling to avoid malnutrition and famine. Selection drives all material progress.

The origins of Darwinism

I love the story of how the idea of natural selection came to light almost as much as I love the idea itself. In the 1830s, both during his long trip around the world and when back in England, Darwin observed the behavior of animals that favored the survival of themselves and their off-spring. For example, when in the Galapagos archipelago in the South Pacific in 1835, Darwin noted that a certain white bird would calmly sit by while the first of its hatchlings killed the second. Why did the bird not intervene—or, if she only wanted only one hatchling, why bother to lay more than one egg? Repeated observation gave Darwin the answer: he determined that a single egg gave only a 50 percent survival rate (survival being defined as that of at least one hatchling), that two eggs raised the survival rate to 70 percent, but that three eggs brought the survival rate below 50 percent. Further, if there were two live hatchlings, the proba-bility of one of them surviving was lower than if there was only one hatchling. Hence the mother's apparently perverse behavior was actually conducive to the survival of her family.

Darwin combined reflections from his field research with two ideas that had been around for many decades in different academic disciplines, and fused them together with explosive effect. The two ideas were com-petition and evolution. Darwin first thought of natural selection in 1838 while reading Thomas Robert Malthus's *Essay on Population*, a dire proph-esy of the effects of competition between individuals for food. Malthus in turn had been influenced by Adam Smith's theories of economic compe-tition in *The Wealth of Nations* (the first volume of which had been pub-lished in 1776). Smith's thinking had been influenced by a writer another century or so earlier, namely the political philosopher Thomas Hobbes, who had in 1651 described society as 'the war of all against all.' So the idea of competition was common currency among intellectuals some 200 years before Darwin published *On the Origin of Species by Means of Natural Selection; or, the Preservation of Favoured Races in the Struggle for Life.*

The idea of evolution had also been widely mooted in the early nine-teenth century. Fossils showed that species had evolved from earlier, more primitive species. K E von Baer (1792–1876) encapsulated a major insight when he stated that 'less general characters are developed from the most general, until the most specialised appear'; evolutionists talked about 'heterogeneity emerging from homogeneity.'[1] What no one before Darwin had explained satisfactorily was how evolution worked.

Natural selection: a simple but subtle theory

Darwin's theory of natural selection is elegant and extremely economical, resting on three plain observations.

First, creatures systematically overproduce their young. 'There is no exception to the rule,' Darwin states, 'that every organic being naturally increases at so high a rate, that if not destroyed, the earth would soon be covered by the progeny of a single pair.' He observes that cod produce millions of eggs. If they all survived, the oceans would be solid cod within six months. Elephants are the slowest breeders of all known animals, yet within five centuries, if unchecked, 'there would be alive fifteen million elephants, descended from the first pair.' Survival is a numbers game, with the odds stacked against most creatures. 'A struggle for existence,' Darwin concludes, 'inevitably follows from the high rate at which all organic beings tend to increase.'

Second, all creatures vary. We are all unique.

Third, the sum of that variation is inherited. We are more like our parents than we are like other people's parents.

Darwin put these three obvious facts together to derive the rudiments of natural selection. Competition among siblings means that only a few can survive. As Darwin wrote with feeling in *On the Origin*:

> all organic beings are exposed to severe competition … Nothing is easier to admit in words the truth of the universal struggle for life, or more difficult— at least I have found it so—than constantly to bear this conclusion in mind. Yet unless it be thoroughly engrained in the mind, I am convinced that the whole economy of nature, with every fact on distribution, rarity, abundance, extinction, and variation, will be dimly seen or quite misunderstood.[2]

Which individual plants and animals will survive? Clearly, those that exploit or fit in best with what Darwin called 'the conditions of life.' In the Introduction to *On the Origin*, he lays out his thesis and acknowledges his debt to Malthus. Darwin comments that he will start by looking at the variation of species, both when domesticated and in nature:

> We shall … discuss what circumstances are most favourable to variation. In the next chapter the Struggle for Existence amongst all organic beings throughout the world, which inevitably follows from their high geometrical powers of increase, will be treated of. This is the doctrine of Malthus, applied to the whole animal and vegetable kingdoms. As many more individuals of each species are born than can possibly survive; and as, consequently, there is a frequently recurring struggle for existence, it follows that any being, if it vary however slightly in any manner profitable to itself … will have a better chance of surviving, and thus be naturally selected. From the strong principle of inheritance, any selected variety will tend to propagate its new and modified form.

Darwin coined the phrase 'natural selection' and explains it very simply:

> This preservation of favourable variations and the rejection of injurious variations, I call Natural Selection.

Plants and animals that have been naturally selected will have had the most successful parents—those who in turn had survived, and came from a long line of survivors—and in turn will have more offspring than other organisms. So in each generation there is improvement, driven by the natural selection of the survivors, and by the relative reproductive success in that generation of the survivors:

> The slightest advantage in one being … over those with which it comes into competition, or better adaptation in however slight a degree to the surrounding physical conditions, will turn the balance.

Darwin keeps hammering home his point that natural selection depends on variation. When the 'conditions of life,' such as climate, change, he says:

> *this would manifestly be favourable to natural selection, by giving a better chance of profitable variations occurring; and unless profitable variations do occur, natural selection can do nothing.*

For most of Darwin's contemporaries, the really controversial aspect of *On the Origin* was not the original part—natural selection—but rather the support that Darwin gave to the general idea of evolution, and especially humanity's descent from animal species. But Darwin's big idea was natural selection. Although he collected (rather inconclusive) data between 1838 and 1859, his main contribution was the flash of insight that he had in 1838: that there was competition for life between individuals and that traits were conserved through their relative adaptability to life's conditions.

Natural selection: the key to life

The process is very simple: variation, then selection, then further variation. Then more variation, more selection, more variation. And so on back to the start of life and forward to eternity. This is how species evolve.

Variation leads to 'better adaption'

Intrinsic to improved congruence with the conditions of life, therefore, is variation. If there were no differences between parents, there would be no differences between offspring. If there were no differences, even between the offspring of the same parents, there would be no basis for differential success. And success is fitting the 'conditions of life.' There will thus be a continual process of improvement or better adaptation to the environment (although, of course, the environment may change, producing different winners and losers).

Variations and improvements occur continually within species, but occasionally a mutation occurs, when an individual has a new characteristic. This mutation may improve or worsen the odds of survival. If the latter, the mutation will die out. If the former, the individual mutant will prosper and leave plenty of offspring, who will inherit and pass on the advantage.

Over time, therefore, most species will evolve positively. And they will respond to any change that the environment brings. When conditions change, new characteristics are required—and encouraged!

For 80 years, scientists have studied intensively one plot of land in the desert in the southwestern United States, photographing its changes in response to climate. They have found that variation is the key to growth. Ecologist Tony Burgess explains:

> *If conditions are variant, the mixture of species increases by two to three orders of magnitude [that is, 20 to 30 times]. If you have a constant pattern, the beautiful desert ecology will almost always collapse into something simpler.*

Diversity leads to efficient use of resources

Darwin suggested that the more species there were on a piece of land, the more efficiently the land would be used. A number of recent experiments have confirmed his hypothesis. For example, research reported in 1984 on 147 plots of Minnesota prairie demonstrated that the greater the number of species in a plot, the more biomass the plot produced, and also the more nitrogen the soil produced; with fewer species, nitrogen leached out of the soil and was wasted.[3]

If a species is diverse, it can survive and prosper. If a species is homogeneous, it is vulnerable.

Take hatchery salmon. In the Pacific Northwest of America, where wild salmon were disappearing, scientists bred huge numbers of hatchery salmon and pushed them into the rivers. But these hatchery salmon had little diversity. They were vulnerable to a slight change in the ecosystem. Too many riverside trees had been cut down for logs. Result: less shade, and therefore a slight rise in river temperatures. Further result: an increase in certain diseases that couldn't flourish in colder water. Final result: the hatchery salmon nearly all died from disease. On reflection, the scientists realised that lack of diversity was the root problem—had the salmon been gradually interbred, allowing mixing and mutation, a diverse adult population would have contained some salmon resistant to the new diseases.

The same applies to computers. More than nine out of ten computers today, like the one I'm working on, have the Windows operating system. These computers have the same core internal components. And every computer with Microsoft software is vulnerable to the same computer viruses.

I don't think it is fanciful to see the same process at work in cities. For example, in 1950s Britain, government mass produced housing for poorer people. Municipal councils built massive tower blocks, all the same shape and pattern (oblong, high, undifferentiated), with all the 'individual' dwellings looking identical. Result: misery, alienation, crime. Something rather similar, although this time produced by private enterprise, was damned in Pete Seegar's song *Little Boxes*, where 'they're all made out of ticky-tacky and they all look just the same.' Jane Jacobs shows in her fascinating book, *The Death and Life of Great American Cities*, that when street lengths, building shapes, sizes, ages and areas within cities are more diverse, then the cities are not only more beautiful, but also more energetic and wealthier.

Diversity works. It always leads to even greater diversity, and to sustainable growth. If we want to sum up the theory of evolution by natural selection in two words, which have great relevance for all societies and businesses, we should simply remember: *diversity works.*

Does evolution imply progress?

According to Darwin, competition and blind chance drive improvement. The struggle for life is at root a lottery, albeit one that may have purpose. Darwin is somewhat ambivalent on this point, but he comments on the dynamics of his 'theory of descent with modification through natural selection':

> *The inhabitants of each successive period in the world's history have beaten their predecessors in the race for life, and are, in so far, higher in the scale of nature; and this may account for that vague and yet ill-defined sentiment, felt by many palaeontologists, that organisation on the whole has progressed … old forms having been supplanted by new and improved forms of life, produced by the laws of variation acting round us, and preserved by Natural Selection.[4]*

Darwin ends his book with an uncharacteristic flourish, designed to make the rather unpalatable notion of natural selection reflect well on the Creator:

> *as natural selection works solely by and for the good of each being, all corporeal and mental endowments will tend to progress toward perfection … produced by laws acting around us. These laws … being Growth with Reproduction; Inheritance which is almost implied by reproduction; Variability from the indirect and direct action of the external conditions of life, and from use and disuse; a Ratio of Increase so high as to lead to a Struggle for Life, and as a consequence to Natural Selection, entailing Divergence of Character and the Extinction of less-improved forms. Thus, from the war of nature, and from famine and death, the most exalted object which we are capable of conceiving, namely the production of the higher animals, directly follows. There is grandeur in this view of life, with its several powers, having been originally breathed into a few forms or one; and that, whilst this planet has gone cycling on according to the fixed law of gravity, from so simple a beginning endless forms most beautiful have been, and are being, evolved.[5]*

Modern biologists are usually extremely careful to stress that there is no implicit evolutionary process leading naturally to improvement; evolution, to scientists, does not imply any immanent purpose or historical progress. Organisms adapt themselves to the conditions of life, but the fact that 'better adapted' organisms thrive at the expense of the 'less adapted' implies no value judgment: better means more rather than superior.

We can choose individually whether, on the one hand, to believe that evolution by natural selection, and the parallel development of humanity's interdependent civilization, where wealth, complexity, specialization and cooperation have all increased over time, are merely happy accidents thrown up by random or indifferent forces; or, on the other hand, to impute some conscious intent or purpose to these developments. Scientists are right not to pronounce on these matters. Yet even if evolution appears to be just a happy accident—which, of course, could in the future turn into a less happy accident—we humans may not be wrong to believe in progress. We can impose a value judgment on pure chance; we can see it as our duty to advance evolution further, even if we do not believe that there was originally any purpose behind it.

Six universal principles implied by evolution by natural selection

In summary, what are the essential patterns that natural selection reveals? Jane Jacobs[6] lays out three themes that were common to all the 'evolutionists' of the nineteenth century:

◆ Differentiation emerging from generality.
One original species leads to all others. New species are formed from an existing species. This is a universal principle: in knowledge, one branch gives rise to one or more new branches, through specialization; in the economy, the same thing happens when one industry gives rise to more specialized branches thereof, or when one firm spawns spin-offs, each of which develops its own particular variations. Variation is the key to development.

◆ Differentiations become generalities from which further differentiations emerge.
Each differentiation becomes a new generality, which can then give birth to new differentiations. Increasing complexity and diversity emerge. As Jane Jacobs says, 'a simple basic process, when repeated and repeated and repeated, produces staggering diversity.' Variation never stops.

◆ Development depends on co-development.

Development of one species requires co-development of other species. 'All forms of life,' Darwin says, 'make together one grand system.' One of the characters in Jane Jacobs' *The Nature of Economies* elaborates on this theme:

> *A horse requires more than its ancestors. A horse implies grass. Grass implies topsoil. Topsoil implies breakup of rocks, development of lichens, worms, beetles, compost-making bacteria, animal droppings—no end of other evolution and lineages besides that of the horse.*

Darwin was alive to nature's web of interdependent species. Today's global economy demonstrates the same pattern of co-development and intricate interdependence.

In addition to these three evolutionary themes, Darwin's theory of natural selection contains another three crucial twists:

◆ The odds against survival are high, leading to a struggle for life.

In nature, in ideas and in economies, so much is produced that only a small fraction can survive. Failure is the normal condition. This implies that only organisms producing many offspring, and generating a stream of new variants, can hope to beat the odds.

◆ The conditions of life determine whether species and individuals survive or not.

In contrast to the French naturalist Jean Lamarck (1744–1829), who claimed that species adapted to the demands of the environment, Darwin held that the environment was the determining factor. For Lamarck, species evolve to survive; for Darwin, species naturally evolve, and the environment decides whether or not they survive.

This may sound a subtle distinction, but it is crucial. Darwin implies that species, and to an even greater degree individuals, cannot hope to control their own destiny. This is a key insight for business, and for individuals in life generally. If a business or a career is failing, there are only two remedies: to change the environment, or to change the character of the business or the individual. Markets usually evolve by changing the winners (whether the 'winners' are firms, technologies or nations), not by changing the way that the incumbent winner (for which read incipient loser) behaves.

In evolution by natural selection, the environment is more powerful than the species, and the species is more important than the individual. In economic development, the market is more important than any particular industry, and the 'species' of producers or consumers is more important than any individual firm or consumer. It follows that if any business enterprise or individual is not being successful, a radical change of environment or behavior is necessary, and even this may not be sufficient.

◆ The process of natural selection contains high degrees of luck, randomness and arbitrary development.

Natural selection is a process of experimentation in which luck is paramount. So is business.

Darwin's economic primer

Bruce Henderson, the founder of the Boston Consulting Group, said: 'Darwin is a better guide to competition than economists.' This is an important observation, although perhaps hardly surprising: Darwin's idea of natural selection was, as we have said, in part analogous to the theories of competition of Thomas Malthus and Adam Smith. So, in applying the lessons of natural selection to business, we are in a sense 'coming home' to a common intellectual heritage. And, to be fair to modern economists, some have incorporated evolutionary aspects into their theories.

Differentiation emerging from generality

The development of economies, industries, individual corporations, and individual careers, follows the same evolutionary path described by Darwin and earlier evolutionists. Differentiation emerges from generality. Specialization means that what was previously one market develops into more than one, ending with the concept of the 'market of one' that can be served economically by tailoring standard product to individual demand.

One industry regularly forks into two: the computer industry, for example, forks into the hardware industry and the software industry. Then the hardware industry forks into the personal computer (PC) industry and the market for larger machines. Then the PC industry forks into the retail store market and the market for direct delivery (via phone and internet). The PC industry also forks into a large number of individual product segments.

This endlessly repeated process results in a richer world. The richest economies are the most diverse, with the greatest degrees of subdivision and specialization. But the process of differentiation from generality also supplies a clue for any business or individual wishing to create a new world that can perhaps be 'owned' by the business or individual. The clue: create a new segment, based on greater specialization. Take one market or industry and make it two. This opportunity always exists—it is the way that markets must evolve, and all that is required to realize a new segment is imagination allied to the following simple method.

Focus on a subgroup of customers that has some homogeneity internally (within the putative customer group) and some differentiation from the rest of the current market. Work out how to serve that customer group better, so that extra value is created for them, but without a proportionate increase in the cost to supply them, and ideally with a decrease in cost. Typically, this can be done by cutting out or downgrading elements of the product or service that are important to the market as a whole but not to your target customer group.[7] Once you have found your new market, use the 'co-development' principle as much as possible—identify techniques and partners from other markets and industries who embody the highest standards of value delivery, who are 'highly evolved' economic species. Then appropriate their ideas or make them partners. Finally, aim to become and remain the standard, the model, and the leader in your new market segment.

So much for general evolutionary principles. But we can go much further: Darwin's particular genius was to build on earlier evolutionary theories, none of which was very specific, and describe the way that natural selection operates. We can now parallel this process in thinking about our own business world. You are invited to collaborate, in a typically Darwinian way, by extending and applying the somewhat general thoughts below to your own more specific, differentiated context, and to take the thoughts to the next level of their evolution.

Where does evolution by selection work in business?

Very clearly, at the level of products. Depending on how good the economic information inside a product is, and how well the product packages the information for customers, and how well adapted the product is to its market, relative to competing products, products will either prosper or die young.

Products live in families, both vertically (over time) and horizontally (at the same time). Products live in a generation game. All products will eventually die, but the most successful products will live long enough to generate at least one 'offspring,' a next-generation product or a same-generation variant. The most successful products will generate many of both. For every successful product, however, there will be many that never got off the drawing board, never survived the test market, died shortly after launch, or never produced any offspring.

Successful products will all be able to say: not one of our progenitors died in infancy. But progenitors are rare. Most products die before they reproduce. Some $160 billion is spent in the developed countries each year on research and development, but only 5 percent of that money succeeds in generating a product or service; and even for the few products that are born, there is a high death rate in the early months and years.

What is success for a technology, a unit of economic information or a product? To beat the odds against selection. To have many offspring. To gain currency: intellectual currency, or dollars.

But what is success for the owners of a technology or of a product? Surely, to make the largest possible long-term profit from selling the technology or product in one form or another. This requires making the product or technology proprietary to the owners. This also tends to produce more variation, more experimentation, more products, more losers and more winners. Recognition of intellectual property rights does not restrict the evolution of economic information; it tends to speed it up, because it encourages variation to justify or to avoid the intellectual property, and to convert it to currency.

Four lessons of economic selection for products and marketing

First, *product ideas will be stronger—more likely to survive and reproduce—if they have emerged from a struggle for life, from substantial competition.* This does not necessarily mean that you should launch a multitude of products, just that you should consider and test a large number, so that those that emerge have faced genuine competition.

For example, the publisher of this book has a policy of only publishing 20 books a year. This is very sensible: it ensures that each book can be properly structured and edited, marketed and promoted. But it would not be wise for him to accept the first 20 books that came along that he felt were acceptable. He should carefully consider perhaps 100–200 potential books, and set up some competitive process between them, before choosing his 20. If he selects the wrong 20, another publisher will gain the market's support.

For some firms and markets, putting out a lot of products and letting the market decide which ones survive *is* a sound procedure. When Sony introduced the Walkman, it flooded the market with hundreds of variants, letting the market decide which few would survive. In credit cards, Capital One is a very successful firm that regularly generates a large number of new ideas, puts them to the market test, and mercilessly kills off the majority that fail. It relies heavily on direct mailshots to attract new customers, puts out perhaps 300 different 'products,' varying the letter, the color of the envelope, the position of the address and so on, and then uses the response rates from its test markets to decide which mailing will be the standard one. The company is currently taking a risk with a major new project: trying to parley its data mining and direct marketing competencies into the sale of mobile phones. If the experiment works, fine; if not, it will be swiftly terminated.

One of the world's most successful consumer goods firms, Procter & Gamble, took the revolutionary and apparently wasteful step, back in 1930, of allowing direct competition between its own brands. This provided challenges that often didn't exist elsewhere in the marketplace. Brands couldn't rest on their laurels; discomfort stimulated improvement and cut complacency. Although this formula worked extremely well, it took almost 30 years for rival firms to copy it. Although it is our best route to success, we hate competition.

Procter & Gamble also has an extremely rigorous and structured product development process, including mandatory test marketing to see whether product concept expectations are met, and whether product sales are sustainable. Few concepts make it through to production. Even successful products are subject to routine, ongoing consumer research, to aid in further product refinement and innovation. P&G has a far higher ratio of potential products to actual products than most of its competitors, and a far greater propensity to produce new generations of successful products.

The second lesson is that *new product variants will arrive sooner or later, whether you introduce them or not*. Natural selection does not care which organism mutates or does not mutate, lives or dies. Economic selection does not care who owns the new product; it just wants to see it arrive. The market doesn't care whether Bic or Wilkinson Sword or Gillette is the market leader in disposable razors, but it does want to see new razors emerge. The market doesn't care whether the big brewers or new specialists supply light beer, or imported beer, or beer from Mexico, or microbrew beer, but it does want to see new eruptions of product variants.

The third lesson, therefore, is to *scatter new breeds around your core product: fill up the potential product spaces so that newcomers can't move into these niches*. You might not think that a new type of product—cherry cola, for example, or healthier versions of mainline food products—has much chance of success. Still, have a go and let the market decide. In the late 1970s and early 1980s it was already clear that 'healthier' food was a growing trend, yet in general the main food manufacturers stood back. The result? New specialists came in and filled the niches. In some cases, as with Wilkinson Sword and Gillette, neglecting to fill a niche can eventually mean a challenge to the core business itself. Fill up all your niches or potential niches. Reinvent your product.

The internet supplies a great example where leading firms have often, or even usually, failed to fill up all the niches or potential niches—and in this case, the internet 'niche' may end up comprising the majority of the market. Whatever else it is, the internet is clearly a channel of distribution that comprises a separate market segment. The leaders in most 'real-world' businesses have been slow to become the leaders in online provision of their service: partly because they have been unfamiliar with the 'virtual' world, and partly because they fear that the internet will simply cannibalize their existing demand, and do so at lower prices. These fears are partly justified, but entirely beside the point. A new niche need not call forth a new leader, but it will unless the existing leader dominates it. Had Barnes and Noble, the leader in real-world bookselling, embraced the internet opportunity at the outset, the word Amazon would still just connote a large river or a nation of female warriors. Now Barnes & Noble is stuck, probably for ever, with having to share its market with the upstart.

The fourth lesson is that *product and service improvement can and should always be accelerated*. Competitive selection drives faster evolution, and evolution proceeds by new and better generations of older products: not just new variants, but also new and better versions of the old product. Tolerate, even encourage failure; it's an intrinsic part of the process. Plan to accept your own failures. Celebrate differences.

Procter & Gamble's experience with Olestra, a fat substitute, is a case in point. The company's original vision was to develop an easily digestible fat that would help premature babies gain weight. The problem was that the fat-like compound, composed of a fat molecule bonded to a sugar molecule, passed through babies unabsorbed. P&G redirected the project to develop a fat-substitute product. Although it took many years to perfect, the process resulted in Olean, a product that tastes and fries like fat but does not digest in the same way. Olean is now incorporated into Frito-Lay chips and fat-free Pringles.

Evolution is slow or non-existent unless there is competition, unless the cycle of struggle for life, selection and improvement or variation is allowed to operate. But evolution can be speeded up: by you or by someone else.

Evolve your products by selection, or someone else may do it for you.

Winners and their sex lives

The few winners should breed prolifically. If the market proclaims a new initiative a stunning success, get the bandwagon going as fast and as far as possible. This means reproduction. It means new generations of improved versions of whatever is successful. It means backing the winners with cash and the best skills available from anywhere.

Too many winners sit in monasteries or nunneries, or opt for vasectomy. They while away their days pleasantly enough, fulfilling existing customers' needs in the same way that they first stumbled across, enjoying easy orders and fat margins. Until someone else invents something new, perhaps an improved version of your own product or service. Winners who don't have sex will die out. Winners who merely have a normal sex life will horribly underperform in relation to their potential. Winners have an evolutionary duty to have a superabundant sex life that produces a large number of offspring.

What does this mean in business? It means taking the winning product or service as far afield geographically as you possibly can—so long as it will still be a winner in the new environment. It means adapting the winner to local markets. It means introducing new generations of product faster and more extensively than those with less successful products do. It means forming spin-off teams and companies that can take what is best and apply it to new products, new customers and new geographic areas. It means squeezing the last particle of possible expansion out of what you have. It means taking a few risks—the risk you should worry most about is the risk of not taking risks.

This is counterintuitive. Surely, it is those who are less successful who should be trying harder to improve what they have? This is normal business reasoning. But economic selection implies that when we have something good and successful, it must be improved and spread, and the new generations improved and spread, as far and as fast as possible. Selection puts enormous pressure for improvement and reproduction on the organisms that are most successful to start with. Selection also gives winners the inbuilt mechanisms needed to keep winning. Use them.

The lessons of selection for firms and divisions

Markets progress via selection. This necessarily requires the *deselection* or death of most of what is tested. This clearly happens at the product and subproduct level. But if there is a free market in companies, deselection will also happen to corporations—they will go bust or be taken over.

In nature, a failed organism becomes food for more successful creatures. The same happens in business. When a company fails, or is taken over, its resources are freed up for more productive use. From society's viewpoint, this is generally a good thing.

Within multibusiness corporations, failing businesses can be tolerated for too long, especially when they are protected from internal or external competition. If a lion reared in a zoo were to escape into the wild, its life expectancy would be low—it wouldn't know how to compete for food. If divisions of large corporations don't have to compete in the market with all comers—for their cash, supplies, technologies, talent and customers—they will rapidly turn into zoo lions. Prolonged subsidy of divisions inhibits evolution.

Fisher's fundamental theorem of natural selection

R A Fisher published his **Fundamental theorem of natural selection** in 1958. It was already known that the average fitness of a population grows from generation to generation. Fisher found that the larger the variance in fitness, the faster the average growth of the population. Greater variation implies greater improvement and therefore faster growth.

Technologies, products, teams, firms and markets that experiment the most and produce most variants will improve fastest, and faster improvement leads to faster market growth. What adapts faster to the conditions of life (if you like, 'what gets better,' according to the judgment of the environment) gains market share and earns superior margins.

Fisher developed the mathematics describing how small variations cause big changes, bigger than one might expect. He showed that if a new allele (an alternative characteristic, explained in Chapter 2), produced by mutation, gives an animal just a 1 percent advantage in fitness, the allele will spread through the entire population within 100 generations. The biological market works quickly and efficiently.

In business, it's difficult to measure something that customers prefer by a margin of just 1 percent. But imagine that there is a 10 percent advantage between one firm's product and another. This will translate into far more than a 10 percent difference in sales, market share and profits.

Pay close attention, therefore, to even so-called marginal customer preferences. Over time, they will count. Where the customers vote you a loser compared to your most significant rival, even if it is by a tiny margin, expect trouble ahead. If you can't persuade a fair sample of customers in your chosen niche that you are measurably better, why are you in the market at all?

Business is a generation game

Markets, products, brands, technologies and companies can be viewed as part of long evolving lines. They are part of a chain of inheritance in which variation is increasing and being tested. The same is true of an individual's career.

Business is a generation game. Each product has forebears and, if it is successful, many generations of offspring. The same is true of markets, of companies, of technologies and of customers. No current generation is an island. It is part of a continent of predecessors and successors. The trick is to find what is most fitted to the environment and ensure that it is passed on and improved as much as possible. Improvement drives expansion.

Fisher's theorem predicts that markets, products, brands, technologies, companies and individuals who improve their fit with the environment faster (than other markets…) will expand faster and be more profitable.

The experience curve—explaining evolutionary improvement

The learning curve, and its derivative the *Experience curve*, help to explain why. The Boston Consulting Group (BCG) found in the 1960s that there was a regular relationship between unit cost and accumulated experience, both for whole markets and for individual firms. As accumulated experience doubles, costs come down by a predictable amount—say by 20 or 30 percent. Accumulated experience means the number of units of something that have ever been produced.

It follows that in fast-growth markets, like semiconductors in the 1960s and 1970s, or software markets in the 1990s, or internet-related markets today, accumulated production multiplies and costs plummet. As costs fall, new applications are opened up. So there is a tautological relationship between cost reduction and growth. Cost reduction is both a cause and a result of growth.

Only by reducing costs or improving features or providing other forms of extra value can markets expand faster than other markets. Above-average growth is the reward for above-average improvement in delivering value.

What happens in whole markets also happens in relation to individual corporations. Firms that grow faster than the market can increase their accumulated production faster than the laggards, and can therefore cut costs or increase value faster. By gaining market share, they actually also underpin the basis for defending and building further market share: they improve their relative cost position or relative value position. It is therefore an excellent strategy—in a profitable market—to build market share even at the expense of short-term profits. This is because by increasing market share the firm will increase its ability to offer customers a better deal. And this is particularly valuable in high-growth markets, because the improvement that is up for grabs is so much greater.

In evolutionary terms, markets, firms, technologies, brands and individuals who gain experience at above-average rates are actually speeding up the evolutionary process. They are packing in more generations in a shorter time. Each generational change offers scope for improvement. Yet improvement only actually occurs if there is adaptive variation; that is, if each succeeding generation or version (of markets, firms, teams) produces something that customers like better, by doing something different that enables the market or firm to deliver better value—and to deliver improvement at a faster and faster rate.

On its own, growth does not necessarily represent success: many firms perversely choose to grow products and businesses that the market does not particularly like. The market typically takes its revenge by ensuring that the growth is profitless. Another caveat is that short-term growth, if not underpinned by the ability to sustain growth by providing real and continuously improving value, may subvert long-term growth or even lead to collapse. For a toy manufacturer to own the Cabbage Patch dolls, or for a publishing house to have a single smash hit title, may be dangerous. There is an optimal short-run growth rate, which may not be the highest available, that will lead to maximum long-term growth—remember Darwin's white bird in the Galapagos, which maximized survival of her family by allowing one hatchling to kill another. But provided that there is innovation, a sustainable ability to add ever-increasing value, the culling of failures and less successful variants, and investment behind whatever the market judges best, growth is the ultimate enabler of business success.

Time-based competition

Time is a proxy for the passing of generations. This idea is closed related to the theory of *Time-based competition*, which states that competitive advantage flows from minimizing the time it takes a firm to complete its activities and provide goods or services to the market, and the time taken to introduce a new product. The higher the time taken, the greater the cost, and the lower the degree of customer satisfaction. Compressing time therefore lowers costs *and* raises market share. A good example is the 'H–Y War' of the early 1980s, which Honda won by introducing new motorcycles to the market faster than Yamaha. First Honda came out with 60 new models in a year, then with another 113 in the next 18 months!

Lifecycles of industries are becoming progressively shorter. Industries that innovate faster grow faster: they gain industry market share, or capture a greater share of an industry value chain. Firms that innovate faster gain share within each market. Individuals who innovate faster grow richer.

Ulam's dilemma—is natural selection unfair?

Mathematician Stanislaw Ulam realized that his discipline spawns nearly 2000 new theorems every year. Very few of these 'survive,' because funding to investigate them is not available. *Ulam's dilemma* was that no one either inside or outside the mathematical profession was qualified to decide which few new theorems should survive and which fall by the wayside. This, he thought, was unsatisfactory:

> *There is no assurance of survival of the fittest, except in the tautological sense that whatever does in fact survive has thereby proved itself fittest, by definition!*

In other words, theories don't win on their objective merits, but as a result of blind competition.

The parallels with business are clear. The 'best' products (or companies, or executives) don't necessarily rise to the top. The standard 'qwerty' typewriter keyboard is far inferior to other forms. Yet I am typing on it now, on a brand new personal computer. English is less efficient than Esperanto. Betamax was probably a better video recorder technology than VHS, which beat it. The market is not always fair, and not always right.

The same is true in nature. Poor early mutations sometimes take hold. Nature is sometimes wasteful and produces unnecessary organs, or allows them to remain (like the human appendix) long after they are redundant.

Poor genes sometimes beat better genes. Sheer luck sometimes means that poor organisms reproduce while their better siblings suffer untimely death. In the short term, at least, nature is not always right.

So what? Over time, selection tends to produce improvement. It is a weird and unfair system, and also imperfect, but on the whole it works extremely well. Almost certainly, no other system would work better.

The implications? Don't buck the market, even if you're sure that it's wrong. If the market feedback on a product or service is negative, trash it. If the capital markets appear foolish or bizarre, put aside your doubts, and try to divine the markets' message for your strategy. Leave judgments on market sanity to investors. Remember the old motto: the customer is always right. So is the market, even when it is 'unfair' or 'wrong.'

Evolving to avoid failure

In Darwin's theory, the 'conditions of life' determine whether or not species survive. The environment disposes of or supports the species; the species does not determine the environment, nor even have the luxury of being able to adapt to it.

In natural selection, failure is endemic. Failure is dominant. Success is the lucky exception.

Is this too fatalistic and severe an insight to be useful in business? In one sense, yes. There are few second chances in nature. But in business there are usually multiple chances.

In a more profound sense, however, the lesson that failure is endemic is invaluable. If we realize that failure is normal, we come to view all business as an experiment, where success at the first attempt is not to be expected. Paradoxically, this is very good news. Whatever our failures, and however bad they are, spectacular success may lurk around the corner—given certain conditions, and with a lot of luck. We would never have heard of Henry Ford if his first attempt at producing automobiles had been his last, or if his second attempt had quenched his determination. His third attempt succeeded, because he finally hit on mass production as a way of lowering costs, and standardization as a way of permitting mass production.

Corporate failure is a good thing. Taiwanese companies frequently fail, yet their backers can easily start again—with superior results to other Asian countries, where unsuccessful companies survive as the 'living dead' because unperforming loans are tolerated.[8]

Natural selection tells us why failure is frequent, and therefore how it can be avoided. Failure implies a lack of fit between the environment and the failure. Natural selection implies that failure is the natural course of events, and that slight modification will not be enough for a turnaround. If a poor fit is to be turned into a good fit, at least one of two things must happen. Either the environment must change, or the failure must change—or, most likely, both must change.

Hoping that the existing environment will change is the prevalent strategy of failing businesses. The strategy nearly always fails, for reasons that Darwin gives. The business has little or no control over the environment, but the reverse is not true. The normal course of events is failure. If failure is to be turned into success, expecting a transformation of the environment is typically futile.

The way out is to find a different environment, and to change the character of the failure. If a company is losing out in one market, it had better find another market more attuned to its capabilities, or change its character radically to serve the existing market in a different way. Since the existing capabilities were evolved to serve the existing market, it is extremely unlikely that there just happens to be another market out there for which the failing business's capabilities are hunky-dory. Variation is therefore probably required along both dimensions. The environment must be different— a different market segment, which implies at least one of the following: different customers, different location, different main business activity, different products or services, different competitors. And the firm's skills must change: different employees, different skill development, different business models, different suppliers, different partners, different standards.

If this attempt fails, fear not. It is time to start another business, using the best parts of the experience, contacts and skills from the existing one.

The same principle applies to individuals. If your career is not making satisfactory progress, it is time to change the environment: get out of your existing business into a new one. But realize also that the 'market'—your previous environment—has given you some valuable feedback. You need to ensure that your character and skills are developed, so that the fit with the new environment will be dramatically better than before.

Summary

Natural selection produces an ever richer, more varied and successful world, relative to the prevailing conditions of life. The results are fantastic. The process is astonishingly simple, but beyond our control and often unpleasant. We must understand the rules of the game, because we cannot change them.

Evolution obeys three principles. Differentiation emerges from generality. Each new differentiation then turns into a generality, from which new differentiations emerge. And development depends on co-development: species evolve more fully when other species co-evolve, and, as Darwin said, 'all forms of life make together one grand system.'

The main business insight from these three principles is the opportunity to create new business segments. This opportunity always exists. All you have to do is to create a more specialized business segment, serving part of the existing environment with a newly evolved business formula and way of working that are particularly relevant to the new segment. In doing so, you should use the 'co-development' principle as much as possible; that is, use the latest evolutions from other markets to help bolster your offering, if possible using new partners who exemplify the highest level of evolution in other markets.

Darwin added the principle of natural selection, where the odds against survival are high, resulting in a struggle for life; where the environment determines whether or not species and individuals survive; and where there is a large element of luck, usually bad luck, endemic in the process.

The principles of natural selection can be observed in business. The only way to win is to be exposed to competition, to pursue variation relentlessly and continuously, to accept the environment's verdict, to be quick—and to be lucky! Products, services, and technologies should emerge from a struggle for life, from some formal or informal competitive process. New products and services will emerge from somewhere: from you or from competitors.

Therefore leave no niche unfilled, no potential niche exposed to innovation from new competitors. Cannibalize your own products and markets before you are cannibalized by others. Product and service improvements can always be accelerated. Realize that failure is par for the course, so encourage failure as a necessary part of the process of experimentation. And when success does come, ride it for all it's worth.

Remember the constant imperatives of life—change; the struggle for life; selection; variation; and further selection and competition. Don't be put off by the odds against success. Where there is life, there is the opportunity to vary; and where there is variation, there is hope of a breakthrough.

Action implications

♦ *Vary, reinvent, multiply, and vary again.* Continuously evolve everything in your business life—your products and services, your ideas, your technologies, your teams, your corporation, your collaborators, yourself. Evolution requires experimentation, variation, rejection of inferior or less well-received variants, hard driving to multiply the successful variants, and a commitment to further cycles of experimentation, variation, selection, and multiplication of the few winners.

♦ *Pursue evolutionary cycles repetitively,* so that the quest for continual improvement via experimentation becomes routine.

♦ *Expose product ideas and products to full-bore competition.*

♦ *Scatter product variants around your core products.* Fill up the potential product space.

♦ *Drive successes as hard and fast as possible.* Ensure that they have prolific sex lives.

♦ *Always accept the market's verdict, even when you consider it unfair.* Don't give new projects or failing products or businesses the benefit of the doubt.

♦ *Only expect to have a few winners—but get the most out of them.* Welcome your failures.

◆ *If you are faced with a failing business or a failing career, quit!* Find a different environment, and change your character to fit the new environment better than any rival.

◆ *Keep reminding yourself that diversity works.*

Notes

1 See Stephen Jay Gould (1977) *Ontogeny and Phylogeny*, Belknap/Harvard, Cambridge, Mass.

2 Charles Darwin (1859) *On the Origin of Species by Means of Natural Selection*, John Murray, London, Chapter III. My quotations are from the 1985 edition from Penguin, London, edited by J W Burrow: see pp 115f.

3 See Adrian Forsyth and Ken Miyata (1984) *Tropical Nature*, Macmillan, New York.

4 Charles Darwin, *On the Origin of Species*, Chapter X, p 342 of the Penguin edition.

5 Ibid., Chapter XIV, pp 459f of the Penguin edition.

6 See Jane Jacobs (2000) *The Nature of Economies*, Random House, New York. This is an excellent short study expressed in didactic dialogue, and I have drawn on many of its themes.

7 This process is called 'value innovation' and is extremely useful, but beyond the scope of this book. For an excellent introduction, see W Chan Kim and Renée Mauborgne (1997) 'Value innovation: the strategic logic of high growth,' *Harvard Business Review*, January–February, pp 103–12.

8 See *The Economist*, January 3, 1998.

2

On Mendel's Genes,
Selfish Genes and Business Genes

It is not success that makes good genes. It is good genes that make success.
Richard Dawkins

People like to think that businesses are built of numbers (as in the 'bottom line'), or forces (as in 'market forces'), or things ('the product'), or even flesh and blood ('our people'). But this is wrong. Businesses are made of ideas— ideas expressed as words.
James Champy

What Darwin couldn't explain

As we saw in Chapter 1, Darwin's explanation of evolution by natural selection was brilliant and convincing: for the first time the mechanism of evolution was fully apparent and plausible. But there was one thing that Darwin couldn't explain. If traits were acquired, how did this work? And how were the acquired traits passed on?

Darwin went round in circles on this issue.[1] In *On the Origin*, he frankly admitted, 'The laws governing inheritance are quite unknown.'

The problem for him lay in everyone's assumption that traits were blended from those of the parents. If this was the case, why didn't individual adaptations become watered down and disappear in a few generations?

Mendel's laws of heredity

From 1856 to 1863—spanning the period when Darwin's great work was published—Gregor Mendel (1822–84), a monk in the Austro-Hungarian empire, experimented with breeding and cross-breeding peas and other plants with distinctive traits. Mendel was surprised to find that the traits of the peas did not blend: a tall plant bred with a dwarf plant led to a tall plant, not a medium-sized one; a yellow pea crossed with a green one did not yield a greenish-yellow pea, but rather a yellow one. When he went on to interbreed the hybrid plants produced by crossing a tall plant with a dwarf plant, although the hybrids were all tall, a quarter of their offspring were dwarf. Mendel correctly concluded that the alternative traits themselves—whether short or tall, this shape or that—were inherited directly, apparently at random.

Mendel's 'law of segregation' states that inherited traits are passed on directly and equally by each parent; rather than blending, the traits remain separate. Each trait is generated by two instructions, with the 'dominant' trait determining appearance, and the 'recessive' trait lying dormant, but capable of emerging in subsequent generations. Mendel also proposed the 'law of independent assortment': which factor is passed on is a function of pure chance, with dominant factors no more likely to 'win' in the next generation than recessive factors. The law of independent assortment also states that individual traits, not the whole complement of characteristics, are passed on in breeding. The seven traits of Mendel's law each operated independently.

Nobody paid much attention to Mendel during his lifetime. Shortly before he died, when he had swapped cross-breeding his peas for less pleasant duties as abbot of his monastery, chromosomes were discovered, although at first nobody knew what they did. The significance of

Mendel's findings only finally became apparent in the 1900s, when it was guessed that chromosomes carried genetic information. Mendel's 'factors' were eventually renamed 'genes,' and it was realized that each pair of chromosomes in a cell carries a great deal of genetic data.

Between 1907 and 1915, American biologist Thomas Hunt Morgan (1866–1945) bred fruit flies, and was surprised to find that one had white eyes rather than the usual red. Even more surprising was that the white eyes were passed on, not in the next generation, but in the one after that; one-third of the fruit flies, all male, had white eyes, exactly as Mendel's laws predicted. In 1915 Morgan wrote *The Mechanism of Mendelian Hereditary*, which showed that genes were physical entities, located alongside chromosomes, and that it was the individual genes that were inherited according to mathematical probabilities.

This finally solved Darwin's dilemma. If inherited traits are not blended, they can be passed on undiluted. Natural selection operates by genes, by genetic inheritance. Morgan also shed new light on how mutations occur: small variations enter the population as 'alleles'—alternative characteristics—with the environment exerting selective pressure on their adaptability. (To use modern language, alleles are different types of genes, like the brown eye and blue eye gene, competing for the same slot on a chromosome; the word allele therefore can be used loosely to mean rival or competitor. Strictly speaking, an allele is 'one of the two alternative positions a gene can have on a chromosome.')

Thus it is possible for there to be considerable variation within one species: mutations do not have to be large jumps. Specific traits can mutate, as well as new species. This is an important conclusion for understanding the nature of business progress.

DNA and its structure

DNA (deoxyribonucleic acid), a large molecule present in the nucleus of every cell of every organism, was discovered in 1869, named in 1899, and largely ignored until the late 1940s, when some scientists began to suspect that it could be the key to how bacteria reproduced. In 1948, the great chemist Linus Pauling used X-rays to work out the shape of proteins, which turned out to be helical, in the form of a helix. In 1953, Francis Crick and James Watson realized from X-rays of DNA that it had a double helix structure, looking a bit like a twisted rope ladder, and in a paper in *Nature* they commented:

> *the specific pairing we have postulated immediately suggests a possible copying mechanism for the genetic material.*

Our genes are made of DNA, a polymer that has a regular, repeating backbone with four kinds of side groups, 'bases,' sticking out at regular intervals. The order of the bases—the way in which the four letters of the DNA language are combined—comprises the genetic message, which can be astonishingly long in a complex organism. Human DNA is thought to contain more than 1,000,000,000 letters. Still, the structure is elegantly simple, and quite universal. All plants and animals share the same basic DNA structure. There are four different kinds of genetic building blocks, connoted *A*, *T*, *C*, and *G*. An *A* building block in a human is absolutely identical to an *A* building block in a butterfly. The difference between people and butterflies lies in the number and sequencing of the building blocks. Every human (except identical twins) has a different DNA sequencing code, yet shares the same structure of DNA with all forms of life.

The discovery of DNA vindicated Darwin's intuition in *On the Origin* nearly a century earlier that:

> *all living things have much in common, in their chemical composition, their germinal vesicles, their cellular structure, and their laws of growth and reproduction … Therefore I infer from analogy that probably all the organic beings which have ever lived on this earth have descended from one primordial form, into which life was first breathed.*

Crick and Watson's discovery also upgraded the significance of genes, leading to 'neo-Darwinian' theories, including the 'selfish gene.'

The selfish gene

Crick and Watson had shown that genes, even internally, are digital. Within a gene, everything is digital code, like a computer language—pure information in digital form. They also showed that the information transfer is irreversible: the gene passes on its information, and the information cannot be supplemented by anything that happens to the body within which the gene sits. Although the gene can be damaged if its vehicle is (for example by toxins or radiation), characteristics thus acquired, like a tan from spending time in the sun, are not passed on to any offspring.

Reflecting on these facts led Oxford biology professor Richard Dawkins to publish *The Selfish Gene* in 1976.[2] Instead of describing natural selection from the individual's angle, Dawkins sees it through the gene's eye. He says:

> *The fundamental unit of selection, and therefore of self-interest, is not the species, not the group, nor even, strictly, the individual. It is the gene, the unit of heredity.*

Here is the gospel according to Dawkins: In the beginning, there were molecules. Then one day, by chance, a remarkable molecule arrived: the replicator. The replicator could make copies of itself. When copies are made, mistakes happen: the copies are sometimes not perfect. The primeval soup therefore began to fill up with several varieties of replicating molecules descended from the same replicator. But the primeval soup wasn't big enough to support all the replicators, so they had to compete with each other: 'there was a struggle for existence among the replicator varieties.' The cunning replicators, the ones that survived, hit on the idea of building survival machines to live in. The survival machines, created by the replicators, got bigger, more varied and more complex. Now the replicators 'swarm in huge colonies, safe inside gigantic lumbering robots;' that is, inside plants and animals. These replicators, now called genes, 'are in you and in me; they created us, body and mind; and their preservation is the ultimate rationale for our existence.'

Natural selection implies the differential survival of entities. Each gene wants to survive, to live a long time, perhaps even to become immortal. The gene survives by making an identical copy of itself; if it can house copies in a long succession of different survival machines (in animals, this means bodies), the gene may survive for a very long time. A gene is potentially near immortal, although it will only survive by collaborating with other genes inside the survival machines. Still, because all the genes can't survive, they are in competition with each other. The gene is 'selfish' because it has been selected only to advance its own cause: to be the survivor in the game of natural selection, where there are always more losers than winners. Genes selfishly compete with their alleles for survival. The genes that survive are the ones best fitted to their environment, which (in a subtle and important twist to Dawkins' argument) includes other genes. In a theme we will come back to in Chapter 5, cooperation turns out to be the highest form of selfishness, both for genes and for their most evolved vehicles, humans.

The theory of memes

Dawkins says that the gene has come to dominate the earth; and the world of the selfish gene is one of savage competition, ruthless exploitation and dastardly deceit. Yet Dawkins does *not* say that our genes control us. Certainly, the genes try to manipulate us, but we can choose to frustrate them, for example by using contraception. Birth rates decline rapidly when the level of education of the women in a society advances.

Moreover, Dawkins offers hope that we can rebel against our genes. Our species is unique, he says, in being able to pass on knowledge in the form of culture: language, customs, art, architecture, science. Humans have invented a new form of replication, a new form of potential immortality, in the form of 'memes,' which is Dawkins' word for units of cultural transmission. A meme might be a book, a play, or an idea—like Darwin's idea of evolution by natural selection. Memes are anything that can be passed on from one person to another, or one generation to another, by means of learning or imitation. As Dawkins explains in *The Selfish Gene*:

Examples of memes are tunes, ideas, catch-phrases, clothes fashions, ways of
making pots or of building arches. Just as genes propagate themselves in the
gene pool by leaping from body to body via sperms or eggs, so memes
propagate themselves in the meme pool by leaping from brain to brain [by]
… imitation.

Dawkins hints that a world of selfishness might conceivably be turned
into something better, if memes with altruistic yet successful features
were to replicate faster than genes.

The idea of memes is controversial. Some biologists do not accept the
parallel with genes, or see the relevance of memes. But to me it makes
perfect sense: memes are a human invention, yet once created they have
a semi-autonomous life of their own; memes replicate, vary, adapt, and
incorporate themselves in robust vehicles; memes produce ever more
complex entities in a way that is very similar to genes.

We should note in passing that there is mounting evidence that
humans are not the only animals where cultural evolution—by which I
mean learned behavior rather than an increasing appreciation of opera—
interacts with and transcends genetic evolution. Research by Dr. Lee Alan
Dugatkin, a biologist at the University of Louisville, shows that even
creatures with low intelligence can imitate the behavior of their peers.
Simple marine bugs called isopods, that are scarcely a quarter-inch long,
have devised a way of copying each other's choice of mates; female guppies (small West Indian fish) will change their minds about which male
guppy to mate with, if they see other females selecting a different male;
and sage grouse shift what they think is sexy according to cultural
idiosyncrasies that vary annually.[3]

A river goes out of Eden

In 1995, Dawkins' book *River Out of Eden*[4] gave us a vivid metaphor for
the spread of DNA. The river of the book's title is of DNA, flowing
through geological time, occasionally branching to form a new species.
Each species' river contains a mass of genes, traveling downstream
together as good companions. 'It is a river of information,' Dawkins says,

that 'passes through our bodies and affects them, but is not affected by them on the way through.' Each river has steep banks, stopping the DNA of one species from getting into another species' DNA river.

Dawkins draws attention to two particular features of natural selection:

♦ Its 'luxuriant diversity.' There are tens of millions of different species. Each species has a different way in which its DNA makes a living, and different ways of 'passing DNA-coded texts on to the future.'
♦ 'Ancestors are rare, descendants are common.' The vast majority of organisms die before they can breed. Only a few of those that do breed will have a descendant alive a thousand generations on. All organisms can therefore look back and say 'none of our ancestors died in infancy,' despite infant deaths being the general rule. It follows that the process of natural selection of genes is extraordinarily discriminating:

Each generation is a sieve: good genes tend to fall through the sieve into the next generation; bad genes tend to end up in bodies that die young or without reproducing … after a thousand generations, the genes that have made it through are likely to be the good ones.

The same selectivity applies to species as a whole. Although there may be about 30 million species on earth, these constitute only 1 percent of the species that have ever lived. Evolution's gate to the kingdom of life is indeed a very narrow one.

The theory of lifelines

The theory of the selfish gene is controversial among biologists. Professor Steven Rose, for example, denounces it as 'ultra-Darwinism' and 'genetic reductionism.'[5] Although that may be unfair, Rose does make a convincing point that evolution happens at many levels, and that the 'lifeline'— the progressive direction of life—includes the evolution not just of genes, but also of organisms and societies. I would add economies to the list; and no doubt Dawkins, if allowed to, would add memes.

The theory of business genes

I would now like to build directly on genetic findings and on Richard Dawkins' theory of memes to arrive at a theory of business genetics that I have called, not unreasonably, the **Theory of Business Genes**.

What is the DNA of business, the most fundamental unit of value? I think that it is 'economic information' and that we may think of units of useful economic information as 'business genes.' Business genes are a type of meme, which, as we've just seen, is Dawkins' word for a unit of cultural transmission. In my definition, a business gene is simply a meme that is related to business, a unit of economic transmission. We could call them 'business memes' instead of 'business genes,' but I have opted for the latter because it makes the parallel with biological genes more explicit.

A characteristic of biological genes is that they tend to travel in packs, and their ability to 'get on' with a large number of other genes is crucial to their success. There are very large numbers of genes present in most animals. Biologists can separate out individual genes, but when it comes to memes or business genes I don't see the point in trying to specify whether we are talking about individual memes/business genes or collections thereof. In practice, most economic information will comprise many different strands or units of information, many individual business genes.

Examples of these business genes or groups of business genes are ideas; the design behind a basic technology such as the steam engine or internal combustion engine, or telephony, or computing; the design for a product component such as the script for a movie, or an integrated circuit; the intellectual capital leading to a piece of software or the kernel thereof; or a formula, such as that for Coca-Cola or for an ethical drug. A business gene is anything intangible that comprises useful economic information and that can be incorporated, alone or alongside other business genes, either into a product or service, or into some vehicle or vehicles that will then provide a product or service.

Business genes are the building blocks of knowhow, of skills and technology in the broadest sense. They comprise economic information that needs to find a commercial vehicle before it can attain its potential and deliver a valuable product or service. Business genes are the origin of economic life. They seek to replicate as widely as possible by incorporating themselves into what we may loosely call commercial vehicles: inanimate things like buildings, machines, software, factories, offices, trucks, and products; but also living things like people, teams, corporations, services, and economies.

Animals and plants are the 'vehicles' for biological genes, the big survival machines into which the genes put themselves. The vehicles do all the hard work to survive and prosper and propagate the genes. The same is true for business genes and their vehicles. The business genes are the invisible business ideas, the knowledge about how to increase wealth; and the vehicles are all the visible apparatus of economic activity: the moving parts, including people, firms, and physical assets, products, and services. The business genes coat themselves with physical texture in order to become more robust, to deliver products and services, and to replicate, just as biological genes need human 'gene machines' for similar purposes. A business gene cannot survive or create value without some physical home; it must be imbedded in something tangible. Even business ideas need some physicality before they can be sold or given away: they must be committed to paper or electronic record, or be communicated from one person to another.

Vehicles are likely to attract good business genes to the extent that they are the best available vehicles for those genes; and they are likely to be successful to the extent that they incorporate the best available genes. The vehicles are the physical expression of economic value and exist to multiply that value. Those best adapted to prevailing economic conditions will flourish; and if the nature of the vehicles, or economic conditions, changes in a way that alters this fit, then the vehicles will cease to flourish.

How business genetics works

Business genes—successful units of economic information—are incorporated into, create, and manipulate many generations of vehicles. These business genes and their vehicles go through a process of evolution by selection, with the struggle for survival, variation, selection, and further variation leading to change, and—on the whole—to improved products and services, and to a richer, more complex, more specialized economy.

Think of how a technology develops: steam power, for example, or nuclear energy, or computing power, or something much simpler and more primitive like the wheel or fire. There are many early versions of these technologies, and they get incorporated into a large number of new products and services. Most of these early technologies and products fall by the wayside: they prove impractical, or too expensive, or else they are supplanted by improved versions of themselves. Any successful technology goes through many generations of experimentation, variation, selection, and improvement. In addition, a successful technology has ancestors that themselves, by definition, survived long enough to give birth to a new technology.

The gene is—depending on which biologist you follow—either the basic mechanism for evolution by natural selection, or the most basic level at which evolution occurs. It is a river of information, flowing not through space but through time. The gene incorporates itself in plants and animals, in machines that are vehicles for its survival. Although they compete with other genes, successful genes have the ability to collaborate with fellow successful genes operating in the same survival machine.

You can see the parallel for technologies or units of economic information or competencies: ways of doing useful economic things. The technologies (and so on) incorporate themselves in products, which evolve because of the competition occurring between the other technologies, and because of the competition between the products themselves. The products with the best business genes tend to survive and multiply and produce better versions of themselves, but what is driving the process is not the products themselves, but the skills and technologies behind them. What survives is not the products, but the useful economic information and technology—the way of doing something useful—that flows into and out of many generations of product. And although technologies compete with each other, as biological genes do, they have to be able to coexist and collaborate with other successful technologies in order to adapt to the environment, just as biological genes must collaborate with many other genes.

And what, for technologies or useful economic information or competencies, constitutes the environment? It is other technologies (and so on), customers and markets, where the business genes have to prove their right to survive by competing for other economic resources.

Business genes in a movie

What, for instance, is the equivalent of the genetic code in a movie? At one level, it may appear to be the digital master recording, from which many copies can be made for movie theaters all around the world. The master recording can then be made into a video, a compact disc, or any other product that can incorporate it. But the digital master recording is itself a vehicle, not a business gene or collection of business genes. The acid test of whether something constitutes a business gene is whether it is information or something tangible: information *constitutes* business genes, but tangible things such as recordings or robots or products or machines are *vehicles* for business genes.

The business genes are the information that goes into the movie and the information that it generates. This includes the skills and reputations of its actors, director, producer and so forth, all of which comprise data of economic value that can be incorporated in future movies and other products. But note that the movie is not the only relevant vehicle for the economic information. The reputations of the actors are carried by the people themselves as well as by the movie. For the same business genes, there are usually several vehicles, of similar types (for example different movies) and different types (for example actors as opposed to movies).

At an even more basic level, the genetic information lies within the script itself. What really matters is not the physical script—the paper or tape or disc on which it is written, which are vehicles—but the ideas it contains. Business genes, generally, are intangibles that have economic value, like stories, customs, ideas, ways of doing things, and the most basic levels of technology: the ideas behind the technology, rather than prototypes or other physical expressions of it.

To qualify as a business gene, three conditions must be met. First, the business gene must be valued, either for its own intrinsic appeal or because it can help to deliver other things that people want, or help to deliver them to a higher standard or with the use of fewer resources. Second, the business gene must be capable of being replicated. Third, it must be intangible.

These three conditions have always been met for the ideas behind successful 'scripts,' like the *Book of Genesis*, or *Romeo and Juliet*, or indeed Newton's *Principia Mathematica*. All of these have been valued; and all have been copied, varied and incorporated into a huge number of derivative products and other vehicles. The genetic code for novels, for instance, is the eight basic plots of which all others are said to be variants.

Take the example of a video on how to sell more effectively, perhaps one made in the 1970s starring John Cleese. After a time the video will die—who nowadays wants to see those flared trousers and kipper ties?—but if the video has been successful, the concept behind it will comprise the real genetic code, the business genes that will live on and be used again in other styles and media.

Humans and business genes

Where do we humans fit in this scheme of things? Can we *be* business genes, or are we always *vehicles* for business genes?

I've suggested three qualifications necessary for a business gene: it must add value; it must be capable of replication; and it must be intangible. Humans can pass the first two tests, but not the third. We can't be business genes. We are never ideas, technologies, ways of doing things, or economic customs. We can create ideas in the first place, and we can take ideas and capitalize on them. Either way, we are vehicles for the ideas and for their replication.

But hang on, I hear you saying, if we can create ideas, create business genes, surely we are their masters rather than their servants. Does this not mean that the parallel with biological genes breaks down? We cannot create biological genes; they create us. Yet surely we can create business genes, and they cannot create us.

If you are saying this, you are right. Remember that what I have called business genes in fact form a subset of memes, Richard Dawkins' term for the social and intellectual replicators that humans have created as an additional evolutionary method to that of genes. Humans create memes; we create business genes.

But humans also stand in a very interesting relation to business genes. We are both their creators, and their vehicles. (The same is true for all memes.) We use business genes, and are used by them. We can propagate business genes that we did not invent. Indeed, this is the normal course of economic progress. For every human inventor of an idea, there can be hundreds or even millions of people who use and develop the idea. Most people who become rich through business do so by using other people's ideas, not their own. Mainly they will elaborate the idea somewhat, so they may create a few minor business genes; but the source of their fortune is mainly the powerful business gene or collection of business genes that they appropriated from elsewhere.

This is precisely the evolutionary process, where a few powerful business genes are replicating very successfully, through a process of variation and continual better adaptation to the environment. We humans are occasional creators of the main business genes, more often the creators of minor business genes, and most frequently of all our role is simply to orchestrate the replication of existing business genes.

Are there different types of vehicle for business genes?

There are two different types of vehicle for business genes: the animate and the inanimate. The former, those that have a life of their own, comprise humans and systems that incorporate human endeavor, including teams, organizations, parts of organizations, cities, and economies. These are all self-organizing systems—a term that we shall meet later in this book—they come together (at least partly) of their own volition and create something that is more than the sum of the parts.

The latter, the inanimate vehicles, include machines, products, physical embodiments of technology like cables and telephone lines, buildings, offices, and trucks, and other 'vehicles' in the usual sense.

The animate, self-organizing systems behave very differently from the inanimate objects. But that is a story for later in this book, and especially for Chapter 9. For our purposes as students of business genes, both the 'live' systems and the inanimate objects are vehicles for the business genes. The only difference is that the human elements contained within the self-organizing systems can create business genes as well as serve as their vehicles.

A new perspective on business

Instead of corporate competition being the focus of study, as in the old economic paradigm, business genetics posits several layers of economic value creation, which is driven by business genes and their struggle for life and reproduction. Humans have multiple roles in the process: as creators of business genes; as users of business genes to create better products and services; and as consumers of products and services and therefore arbiters of their survival, spread, and demise. Corporations are important intermediate vehicles, but there are many other types of vehicle; and corporations derive their power from being the best vehicles for business genes and for their creators, the entrepreneurs and knowledge workers.

Small companies go bust all the time, in large numbers. Although this is painful, it is a necessary part of economic progress. The companies that are better suited to the environment—which largely means the market—survive, and are stronger for having to face competition. As the great economist Joseph Schumpeter told us in 1942, capitalism progresses via a process of 'creative destruction.'[6] When a company is destroyed, it frees up resources for better use elsewhere.

There is nothing sacrosanct about corporations. As vehicles, they are only useful to the extent that they are the best possible incarnation of business energy and information. If that business energy and information would be better deployed elsewhere, we should throw away the old vehicle and use or create another one.

Large companies are less likely to go bust than small ones. They have become large companies by being one of a small minority of small companies that have been very successful; the large companies have undergone a long process of selection for the privilege of being big. Yet large companies can set the cause of business evolution back if they use their size—as many do—to insulate themselves from competition. This may work for a time, but natural selection says that insulation from competition halts or at least slows down improvement in products and services (common sense and observation tell us the same). Selection also tells us that someone, somewhere, will be experimenting with new products or technologies that may eventually become a challenge to the sclerotic old firm. The reckoning can be postponed, but not averted, and when the firm eventually faces competition it may quickly crumble. Think of the near collapse of IBM. Think of the actual collapse of the Soviet Union, which at one time must have been, among other things, the world's largest corporation.

The process of killing off firms that have outlived their usefulness will be accelerated if we adopt the business gene view of corporations. Vehicles that are no longer working well should be abandoned by healthy business genes—and sooner rather than later.

Spin-offs

Spin-offs are growing in popularity, but are still not used enough. I use the term 'spin-off' loosely, to mean not just firms that are still owned by the proprietors of the original firm, but also those where the key players come from a common 'parent,' and bring with them important parts of its knowledge or customer base.

Is it a coincidence that industries where there are many spin-offs also tend to grow faster than other industries—and also deliver faster increases in value to customers and investors? Think high-tech. Silicon Valley is stuffed full of spin-offs. Think management consulting. Think venture capital. Think investment banking. Spin-offs and 'team moves'—a sort of half-way house to a spin-off—are endemic in such industries.

In boring industries, where progress comes in arthritic snail paces, spin-offs are rare. Perhaps if there were more spin-offs, these industries would become more interesting and successful.

Natural selection predicts what happens with spin-offs. The new firm takes with it much that is good from the old firm—it inherits the latter's good genes—but it also adds new twists. If the innovations suit the market, the spin-off prospers. Spin-offs from the successful spin-off then repeat the process. No one wants to spin off from an unsuccessful firm.

Typically, the owners of successful firms do not benefit from spin-offs. The children go their own way and the parent gets no benefit. How much better for the owners of the successful firm if they can have a share of the spin-off. That this does not happen more often is due to a lack of foresight, combined with managers' natural reluctance to risk the creation of new, independent units that will not be under their control. Yet natural selection points to a different model. Every successful firm should have many spin-offs. Owners who act in time to back promising spin-off ideas, and if necessary overrule their managers, can capture a share of the value created by the spin-off.

If you want to give nature or the market a helping hand, take part in a spin-off.

Now let's complete our theory of business genes by looking at one final piece of genetic research, dealing with the problem of inbreeding.

The Hardy-Weinberg law

The study of genes has shown why taboos on incest are well founded. We carry harmful alleles that do not get expressed because they are 'recessive' and occur in only one copy. The odds against two similar alleles fusing are very high. But inbreeding cuts the odds against this misfortune considerably.

Also of interest is the **Hardy-Weinberg law**, discovered independently in 1908 by British mathematician G H Hardy and J H Weinberg, a German physician. The law says that the probability of inheriting a particular allele from a parent is exactly as high as that of passing it on to a child. The allele frequencies do not change.

Inbreeding and the Hardy-Weinberg law are useful concepts in thinking about how evolution is restricted, in a technology, product, company, market, or nation, when there is no significant change in its 'gene pool.'

Take the case of an organization. The 'gene pool' is not just the skills of the senior executives. It comprises all the important inputs to the firm, including what suppliers contribute, what technologies and distribution channels are used, how customers are used to help the firm improve, what investments are utilized, all collaborative networks (which are not confined to suppliers—important collaborators for high-tech firms, for example, include individuals and teams within universities, who may not have any contractual relationship), and all employees and new recruits to the firm. 'Inbreeding' occurs when this total gene pool is not sufficiently replenished, changed, and stirred. Options for change should continually be evaluated.

Business genetics for executives— six action rules

◆ Use the best business genes available

There are three ways to deploy business genes. One is to create them from scratch: invent a new product or service, or a new business system. This is a rare event, and few of us have the originality required. A second way is to appropriate and use successful business genes. Remember, the genes want to multiply, so they are amenable to being used. But to do this successfully you may need to get there ahead of other people who might have the same idea. For instance, my friend Raymond Ackerman, who created the Pick 'n Pay supermarket giant in South Africa, used the idea—the business gene—of self-service supermarkets after it was already well established in the US but before it had become so in Africa. A third way is to take an already successful business gene and modify it slightly: create a new variant of the business gene.

◆ Make yourself an excellent vehicle for successful business genes

To use successful genes, you must further their purpose and help them multiply. This requires adaptation on your part. Adaptation requires exposure to competition. Don't seek to insulate yourself from internal or external career competition; if you do, you'll stop developing. Compete in the major markets, not in the backwaters. Beware of working in corporate pyramids, which insulate executives, especially senior ones.

◆ Use the best vehicles available and drive them

You are a user of economic information, you are the skills that you have, including the skill to collaborate with successful business genes and other vehicles for them: with other individuals, with teams, and with organizations. You are the value added. You are the driving force. The team or company you join, the other resources you commandeer, are vehicles for you. The vehicles are there to advance your purpose, to provide protection, to incarnate your energy. Remember that the vehicles are just that, and the only reason to work through one is if it is *the best possible vehicle around for your purposes*. Continually ask yourself: Am I driving or being driven? Am I driving the right vehicle? Is there anywhere else I could add more value?

◆ Career evolution requires variation: a series of new jobs (whether or not you change firms) and new ways of doing the existing job

Start a project. Take on a new responsibility. Change the furniture. Identify new business genes that can provide you with fresh direction and for which you can be the best vehicle.

◆ Evolution requires continual experimentation and improvement

If you don't produce a new-generation version of yourself as quickly as your career competitors do, you'll fall behind. Experimentation and improvement require fresh combinations of new business genes—new skills, new ideas, new ways of working. Remember that your success requires you to be a vehicle for excellent combinations of business genes, and to work within or alongside other successful vehicles, so experiment all the time with new combinations of genes and vehicles.

◆ Evolution requires failure

The greatest and most abundant freedom that the universe offers is the freedom to fail. For most organisms this means an early death, and it is the species that benefits. Fortunately, in careers we always get another chance. But constructive mutation in your character and skills requires failure, as well as the maturity to recognize and accept this. Don't let your ego deny the fact of failure; this is another form of competitive insulation.

Accepting failure and using self-development processes to benefit from it can free you to compete. Failure is easier to accept and to reverse in the context of business gene theory. It results from being a good vehicle for poor business genes, or from being a poor vehicle for good business genes. Identify which is the case, and take appropriate corrective action.

Summary

Mendel's laws, Thomas Hunt Morgan's experiments, and the discovery of genes showed how inheritance operated, and added a micro dimension to evolution. Specific traits can mutate, not just new species. Crick and Watson's discovery of DNA and its structure revealed the common links between organisms, and the fact that evolution is driven by passing on information. Richard Dawkins' brilliant theory of the selfish gene gives a new perspective on evolution, where the gene is the driving force and organisms are merely vehicles for genes. Dawkins' theory of memes proposes that humanity has developed a new form of replication: units of cultural transmission, information that can leap from brain to brain, and can supplement or even compete with selfish genes.

I have proposed a parallel theory of business genetics, based entirely on insights from Mendel, Hunt Morgan, Crick and Watson, and Dawkins. Business genes are basic units of economic information, intangible ideas of value that can be replicated to create yet more value. These are what ultimately drives improvement. Business genes are created and used by individuals and teams; but once created, the business genes have a life of their own. They seek to incorporate themselves into as many vehicles as possible in order to spread themselves widely. The vehicles fall into two categories: machine-like objects; and self-organizing systems, including individual executives, teams, and organizations. The vehicles produce products and services, which themselves thereby also incorporate business genes.

At all levels—the business genes, the vehicles, and the products—the same process of evolution by selection operates. Business genes survive and spread by being the best of their type; the process spawns many new variants and many dead, redundant business ones. The vehicles survive and prosper by being the best vehicles for the best business genes; they will be cast aside by the business genes when this is no longer true. The products and services are produced by the vehicles and incorporate the best business genes; the products and services also face a struggle for existence, with new and improved versions continually replacing earlier ones.

This view of business moves the center of gravity from the corporation back to the fundamental sources of value: to business genes, to the genetic code of business, which is useful economic information. Fundamental value does not reside in products and services, nor yet in the vehicles that produce them—corporations and the physical things that they own—but in the ideas, formulae and technologies that drive the vehicles. The best business genes will always seek out the best vehicles, and if a better vehicles becomes available, the business gene will drive that vehicle, leaving the earlier vehicle stalled.

To create value, individuals may create new business genes, although these genes are slippery and impossible to control—you must also create and keep the best vehicle for the genes, or they will go elsewhere. However, value may also be created—rather more easily—not by creating new business genes, but by identifying valuable business genes that are underexploited, because the appropriate vehicles for them have not yet been created. Genes need vehicles; they can identify the best available, but cannot necessarily create the vehicles themselves—and in business, this results in opportunities for individuals to create appropriate vehicles. Value can also be created by combining business genes in new permutations, and providing new vehicles for the new mix of genes.

Action implications

◆ *Identify business genes that are underexploited and that are having to make do with poor vehicles.* Create the best new vehicles for existing powerful business genes—for valuable economic information, ways of working, and technologies.

◆ *Join existing successful business genes into new combinations,* and provide appropriate vehicles for the new combinations.

- ◆ *Make yourself the best vehicle for a winning and unique combination of business genes.*
- ◆ If you participate in running an organization, *realize that its value derives from being a vehicle for successful business genes.* Ensure that the organization is and remains the best vehicle for the genes. See that the organization's gene pool is continually replenished with new inputs that have already proven their success.
- ◆ *Create spin-offs and spin-outs from existing organizations.*

Notes

1   Darwin eventually settled on the (wrong) idea that cells throughout the body contribute instructions to the reproductive cells, thus enabling traits to be passed on to offspring.

2   Richard Dawkins (1976, revised edition 1989) *The Selfish Gene*, Oxford University Press, Oxford.

3   See Lee Alan Dugatkin (1998) *Cheating Monkeys and Citizen Bees: The Nature of Co-operation in Animals and Humans*, Free Press, New York; Lee Alan Dugatkin and Jean-Guy J Godin (1998) 'How females choose their mates,' *Scientific American*, April 1998, pp 56–61, and Lee Alan Dugatkin (forthcoming) *Guppy Love: Genes, Culture and the Science of Mate Choice*, Free Press, New York.

4   Richard Dawkins (1995) *River Out of Eden*, Weidenfeld & Nicholson, London.

5   Steven Rose (1997) *Lifelines*, Allen Lane/The Penguin Press, London.

6   Joseph A Schumpeter (1942) *Capitalism, Socialism and Democracy*, New York, Harper & Row.

3

On Gause's Laws

Competitive strategy is about being different. It means deliberately choosing a different set of activities to deliver a unique mix of value.

Michael Porter

Gause's principle of survival by differentiation

Soviet scientist G F Gause did some very interesting experiments on small organisms. He put two protozoans of the same family but different species in a glass jar with limited food. The little creatures managed to cooperate and share the food, and they both survived.

Then Gause put two organisms of the same species in the jar, with the same amount of food as before. This time, they fought and died.

I call this **Gause's Principle of Survival by Differentiation**, or PSD for short. Later in this chapter I'll explain why I think the PSD is so important, but for the moment, trust me. Try to keep Gause's organisms in your mind's eye, because if there is one image I'd like you to take away from this whole book, it's probably that of Gause's protozoans and the PSD.

Incidentally, Darwin anticipated the results of Gause's experiment in Chapter III of *On the Origin of Species by Means of Natural Selection*:

the struggle [for existence] almost invariably will be the most severe between the individuals of the same species, for they frequent the same districts, require the same food, and are exposed to the same dangers ... As species of the same genus have usually ... some similarity in habits and constitution, and always in structure, the struggle will generally be more severe between species of the same genus, when they come into competition with each other, than between species of distinct genera... We can dimly see why the competition should be most severe between allied forms, which fill nearly the same place in the economy of nature.

Gause's principle of competitive exclusion

Gause's experiments also led to the conclusion that two competing species can only coexist if there is more than one scarce resource.

Two populations compete if one lowers the growth rate of the other. They may do this by eating each other's lunch, crowding each other's space, or playing their Walkmans so loud that the other population commits suicide. Whatever.

Coexistence, dominance, and bi-stability

Gause found three outcomes from the protozoan wars:

◆ Two species both invade each other equally. The boundaries between them break down; they end up *coexisting* in the same space.
◆ Only one of the species invades the other. It ends up *dominant*. The invaded species is wiped out.
◆ Neither species invades the other. Like the late, lamented arms race, there is a balance of power that ensures peace. Biologists call this *bi-stability*.

Ecological niches and MacArthur's warblers

In ecology, a niche is not just a place where a particular creature lives; it is also a way of making a living—a special way of obtaining resources, a specialized job, or, in Darwin's happy words, 'a place in the economy of nature.' Ecologist Robert MacArthur has shown that for warblers, a spruce tree is not a spruce tree; it is actually several different niches for several different types of warblers. Apparently, each type of warbler has its own bit of the tree to which it does different things. Each warbler has its own small and specialized ecological niche.[1]

It turns out to be quite difficult for ecologists to map the boundaries between ecological niches, but the principle is clear and very useful: *each separate niche sustains just one specialized type of plant or animal.* Note that the specializations are carried to a very high degree and that each creature does just one thing in one place. The warblers in the forests of the northeastern United States are strong believers in Adam Smith's division of labor, the principle of specialization allowing higher productivity.

Finding *unique* niches

The first lesson from Gause's Principle of Survival by Differentiation is that you want your competitors to be at least slightly different from you. They can be of the same family, but not of the same species. Now, remember that there are perhaps 30 million species on earth, so making your firm a separate species, if it isn't one already, shouldn't be impossible. If two organisms of the same species compete in the same space with a limited market (food), they fight and die. If they are different, they can cooperate and both live.

MacArthur's warblers found different parts of the spruce tree because they were different types of warbler. They had found unique niches, each warbler with its own bit of the tree.

So *your firm needs unique niches, places where no one else can go because they aren't exactly like you.* The 'places' can be particular customers, geographic markets, channels of distribution, products, technologies, or any other source of differentiation, but at least one of these must be unique to your firm. Otherwise you are the same species as your competitor, slugging it out in the market equivalent of Gause's glass jar.

Bruce Henderson, who found his unique niche as founder of 'intellectual' strategy consulting and who was a keen student of biology, put it well:

> *Competitors who prosper will have unique advantages over any and all competitors in specific combinations of time, place, products and customers. Differences between competitors is the prerequisite for survival in natural competition. These differences may not be obvious. But competitors who make their living in exactly the same way in the same place at the same time [won't prosper].*[2]

Who can invade whom?

Gause's Principle of Competitive Exclusion highlights the symmetry or asymmetry in any competitive struggle. Your firm is in a very weak position if your rival can enter your space, but you can't enter his. (This, incidentally, is the organizing idea behind the 'world domination' board game Risk.) If you are in this position, don't stay there! You must find a niche, a way of earning your living in a different way. If you can't do this, and the business is still profitable, sell it before it is too late!

Look at what happened to the British motorcycle industry in the 1970s. Honda had two unique markets that the British motorbike makers couldn't enter: it had the large Japanese market, and the market for small bikes (clearly these overlap, but they are conceptually separate and, outside Japan, physically separate too). At first, the British had a separate market: that for larger bikes. But the relationship between the British and Honda was asymmetrical. The British couldn't enter Honda's markets, because British bikes were designed to be big and powerful, and were too expensive to downscale. But Japanese bikes were designed to use modular components, and to be capable of upscaling. Honda's bikes were also cheaper and better value, even after the cost of transport and distribution, which was another reason for the asymmetry.

Before the Japanese entered the British bike market, the British industry might have felt secure. Even after the Japanese entered, did it matter? They were only selling 'toy' bikes, often to people who'd never bought a bike before. If the British bike makers had read about Gause's protozoans, they would have known their fate. Only by developing a unique niche which the Japanese couldn't enter—as BMW did with its ultra-comfortable bikes for big bottoms—would the British have survived. By 1980 their bike industry was all but dead.

Of course, *if your firm can invade and the competitor can't, you should.* You can feel very confident of the outcome.

Bi-stability is better than coexistence

Gause's research makes the interesting contrast between bi-stability and coexistence. Coexistence is real competition, where either can invade the other. Bi-stability is when they can't; it's illusory competition.

Bi-stability implies that two populations are not really competitors. They are both excluded from each other's domain. This is very common in business. An industry may appear to be competitive, yet each competitor has different customers, different distribution channels, or some other differentiating factor that has barriers around it. Within each segment, different firms can enjoy high market shares and high profits. Whenever you see an industry that is very profitable, like high-end consulting, you find a market where competitors are very specialized and where each has a high share of its own niche. Smart players will keep it that way.

On the other hand, a market characterized by many coexisting competitors is likely to be one in competitive stalemate. No one will have the edge. Market share will have little value. Here, the struggle between competitors for customers makes every supplier a loser.

The only hope within a coexistence game is to recognize it for what it is, and for all the competitors to signal to each other that they will live and let live. Price wars, aggressive marketing campaigns, or indeed vigorous activity of any kind: all of these must be avoided like the plague. To break out of a coexistence game requires the development of a segment where dominance can be exercised—and we are back to the fundamental point, that you need to be different.

Breaking out of coexistence and into bi-stability

If you're in a coexistence market, it may be possible to break out and make your own 'bi-stable' market.

The airline industry is, in general, a good example of coexistence. Except where regulation restricts the number of competitors on any route, consistent profits are hard to come by. Without regulation, each airline can invade the others' customer base. Each innovation, such as free drinks in the cheapest class, can easily be copied. Only fare structures of byzantine complexity keep the airlines from making losses; and fares are becoming increasingly transparent as sophisticated customers use the internet to search for the lowest fares. Yet even in the airline industry, it's possible to escape coexistence by doing something radically different.

Take Southwest Airlines. It doesn't follow the typical 'hub and spoke' system based around large airports. It avoids long routes, in-flight feeding, and ticketing bags to other destinations. It offers one-class, frequent flights between a few, carefully selected cities; short check-in times; automated ticketing; and low fares. It has a standard fleet of 737s, cutting maintenance costs and delays. It encourages direct payment, cutting out commissions to travel agents. It appeals to a particular type of traveler who appreciates the trade-offs that it makes.

Other airlines can't copy Southwest: there's not room for two such airlines on any of its routes, and the competitors are not set up for this approach anyway. Southwest has therefore created a 'bi-stable' market, and can enjoy high profits despite having low prices. The trick is to have higher utilization rates than is possible under coexistence.

The danger of having one key to success in an industry

Gause's test-tube wars revealed one further fascinating fact. If scarcity in one single resource like food *or* air was the issue, one species always became dominant. Coexistence or bi-stability only worked if more than one limiting factor was relevant.

The corporate equivalent? When competition only proceeds via one variable. If price is all that matters, the low-cost corporation is bound to win. If quality reigns supreme, the supplier perceived to have the highest quality will dominate. If innovation is all that matters, as in some fashion markets, then the trendiest will inherit the earth.

The punch line? If you're in a market where there is only one dimension of competition, and you're not the best on that dimension, you should create a separate segment where another dimension matters. If you can't do this, and the market is unprofitable for you (as it probably will be, if you're measuring the profitability properly), then you should quit.

Differentiation is a set of actions to take you to Uniqueland

Differentiation is not quite the same as specialization. It is actually more fundamental. Differentiation implies that you don't just specialize; to differentiate successfully, *you find the market space where the ratio of your productivity to that of your best rival in that space is highest*, relative to all the other areas where you could specialize; and then *make the way you earn your living as different as possible from that of anyone else*. You actually accentuate the differences between you and the nearest competitor until you have arrived at a unique niche.

Differentiation is not a description of a state; it is a set of actions to increase differences.

Think how you could further differentiate your firm from its closest competitor (in each area where you face a different 'closest competitor') along each of the following four dimensions:

◆ customer type
◆ type of product or service
◆ geography
◆ stage of value added.

Extend the distance between yourself and your closest rival, until it is
clear that there really *isn't* a closest rival. Then you've arrived at the happy
country: Uniqueland.

Summary

Gause's Principle of Survival by Differentiation tells us that we cannot
expect to prosper if we make our living in the same way at the same time
in the same place as another similar competitor.

Gause's Principle of Competitive Exclusion says that if one species can
invade another, but the other can't reciprocate, then the former will end
up dominant. Therefore you should not enter or stay in businesses where
you can't enter some of your competitor's markets but it can enter all of
yours. On the other hand, if you can invade a rival who can't invade you,
go ahead.

Competitive exclusion also implies that spurious or illusory competition can exist, where each player is excluded from the other's domain.
This is called bi-stability and is usually a happy situation. Not so good is
coexistence, where each player could invade the other.

Gause's third and final message is that when a single resource is of
paramount importance, one species—the one best at cornering that
resource—will end up dominant, and the other species will be wiped
out. There is a perfect analogy in business, because in some markets only
one purchase criterion matters, and the firm that is best at meeting that
criterion will dominate. If you are in such a market and are not best at
what is most important to customers, either find a segment where the
main purchase criterion is different, and you can meet it best, or else quit.

The idea of ecological niches says that there is an amazingly large
number of ways of making a living, and that the ideal position is to have

your own niche where no one else makes the same living in the same way at the same time. Then there are no competitors. You are unique.

Action implications

◆ *Differentiate*. Specialize. Put all your energy and resources into areas where you are substantially different from any rival.

◆ *Systematically increase the degree of difference from your closest rival*, along the four dimensions of product/service type, customer type, geographic market, and stage of value added. Aim to become unique; that is, to have no rivals.

◆ *Get out of businesses where a competitor can invade you but you can't invade it.*

◆ *If you can invade a competitor, but it can't invade you, go ahead and invade.*

◆ *Avoid businesses where there is coexistence*: where everyone can snatch each other's customers. If you're in a coexistence market, turn it into a bi-stable one by finding a distinctive approach that customers like and competitors can't imitate. If you can't do this, and are stuck with coexistence, make the most of it by trying not to step on competitors' toes.

◆ *In any business where a single purchase criterion (such as price, or quality, or service) is all that matters, and you are the best at satisfying that criterion, go all out for dominance.* Widen the gap between yourself and competitors on that criterion. If you are in such a business, yet are not the best at meeting the key criterion and cannot become the best at it, then find another segment (within or outside your current segment) where the purchase criterion plays to your strength. If you can't find such a segment, exit.

Notes

1   R H MacArthur (1958) 'Popular ecology of some warblers of northeastern coniferous forests,' *Ecology*, 39, pp 599–619.

2   Bruce Henderson in Carl W Stern and George Stalk Jr. (1998) *Perspectives on Strategy*, John Wiley and Sons, New York.

4

On Evolutionary Psychology

Behind every successful man stands a surprised woman.
Maryon Pearson, wife of Canadian Prime Minister

Meet the Flintstones—in the office

One fascinating new scientific discipline that evolved in the last third of the twentieth century is ***Evolutionary Psychology***, a mix of genetics, anthropology, palaeontology, neuropsychology, and social psychology. A form of neo-Darwinism applied to the study of human behavior, evolutionary psychology has some controversial but fascinating messages for business.

Punctuated equilibrium

In 1972, two evolutionary biologists, Stephen Jay Gould of Harvard University and Niles Eldredge of the American Museum of Natural History, proposed the idea of ***Punctuated Equilibrium***. This is a theory that we will examine in more detail in Chapter 11, but it's useful to introduce it here because it helps us to understand why our genes may lag behind changes in society. The theory of punctuated equilibrium says that evolution proceeds by long periods of relative quiescence and stability, punctuated by short periods of rapid change.

The disappearance of the dinosaurs marks such a punctuation point. They dominated the planet for 130 million years, then zap—they either disappeared when a comet hit the earth and set off eruptions of sulphuric volcanoes that left a persistent cloud of sulphuric dust; or, perhaps more plausibly, they gradually evolved into birds. Either way, it was quite a change!

Man's infrequent punctuations and the theory of evolutionary psychology

Business, society and technology seem to operate via punctuated equilibrium, although on a different timescale. Karl Marx divided history into three phases—feudalism, capitalism and socialism—and hypothesized that the transition from the second to the third would be rapid. More conventional historians point to just two punctuation points during the whole of human history: the transition from hunter-gatherer Stone Age society to that of agriculture, and the relatively recent shift to an urban, industrial society.

The basic thesis goes like this. Humans emerged as hunter-gatherers, living in clans, about 200,000 years ago, and evolved traits suited to that life. Then, a mere 7,000 years ago, agriculture developed, leading to a totally different society. A little over 200 years ago, industry and commerce began to prevail over agriculture. The conditions of human life were utterly transformed. Before the eighteenth century, nearly everyone lived on the land, with malnutrition a constant threat. The main sources of power were the horse and cart, and human brawn. We forget the extremity and recency of our shift to an urban and generally prosperous society that harnesses machine and brain power.

So our conditions of life have changed radically since the Stone Age. But we haven't. Evolutionary psychologists contend that we are still 'hardwired' with the same circuits that were functional for clan-living hunter-gatherers. We are geared for living in the Stone Age. Why? Because 7,000 years (and certainly 200-odd years) is just not long enough for human evolution to produce genetic traits that match the new sur-roundings. As Edmond O Wilson comments:

> *The culture of the Kalahari hunter-gatherers is very distinct from that of [modern-day] Parisians, but the differences are primarily a result of diver-gence in history and environment, and are not genetic in origin.*

In what ways are we still prisoners of the Stone Age? Evolutionary psy-chologists say that our emotions take precedence over our reason. We incorrigibly exhibit primitive behavior: we go by first impressions; beat our breasts; develop clans; dislike outsiders; follow the herd; gossip; con-struct informal hierarchies; follow confident leaders unthinkingly; and (I said it was controversial) live up to sexual stereotypes. We may know that these behaviors are primitive, and often harmful to ourselves and those around us. We may understand, intellectually, that these actions belong to the Stone Age, not to the global world of twenty-first-century business. But we just can't help ourselves. We're hardwired to be what our genes make us: Stone Age animals.[1]

Many scientists scorn evolutionary psychology, yet it already has an impressive body of evidence behind it.[2] The points that it makes about business life are very resonant and useful.

The neurology of Stone Age man

We may summarise the apparently immutable characteristics of humans, according to evolutionary psychology, under four headings:

◆ Dominance of emotion over reason.
◆ Predictably primitive behavior.
◆ Avoidance of risk.
◆ Mad scrambling when under serious threat.

The dominance of emotion over reason

Alert instincts were literally vital on the Savannah Plain, and later, during the Ice Age, on the mammoth steppe. Hunter–gatherers were exposed to predatory short-faced bears, sabre-toothed cats, lions, and wolves; enemies that might raid them; and weather that might flood, freeze or bake them. Good instincts saved lives. Those with good instincts passed on more of their genes and instincts became honed for survival. Emotions were and are the first reaction to everything seen or sensed.

So, evolutionary psychologists contend, when we receive feedback, especially if it has a negative element, our natural disposition is not to think about it. Instead, we react emotionally. We react to people the same way. Do I like her? Not: Is she valuable to the company? Nor even: Can she be useful to me?

We are easily manipulated, even when we realize it, because warm emotion is so much more important to us than cool reason.

Predictably primitive behavior

Cooperation, sharing, specialization and friendliness
One highly functional aspect of the hunter-gatherer was the disposition to live in fairly large groups and to cooperate with the rest of the clan. *Homo sapiens* emerged as a highly social animal, a bit like hyenas and lions, only much more so. Successful Stone Age people lived in large clans, containing up to 150 people, according to Robin Dunbar, a psychologist at the University of Liverpool.[3] Professor Dunbar also tells us that the larger the troop size in primates, the larger the brain. Anthropologists have found that, even among primitive societies, the richer and more developed ones have a greater number of roles.

Sharing food was the basis for cooperative exchange among hunter-gatherers. The humans who survived and prospered and passed on their genes were skilled at cooperating peacefully and at exchange. The division of labor appears to have occurred right back in the Stone Age: one man would make spears, another knew how to hunt game, a third was an expert spear thrower. This supposition is supported by evidence about modern hunter-gatherers: for example among the Ache of Paraguay, who are still hunter-gatherers, some men specialize in finding armadillos in their burrows, others in digging them out.

Virtually all hunter-gatherer tribes exhibit sexual specialization: men hunt, women gather. According to Matt Ridley, 'Women are more verbal, observant, meticulous and industrious, skills that suit gathering.'[4] (It has been suggested, however, that humans also benefit from substantial similarity between men and women, allowing cooperation and substitution of tasks, an evolutionary advantage denied to species such as spiders where males and females are dissimilar, or moose or walruses where the genders lead largely separate lives.)

The animals hunted were big, and successful hunts might be infrequent. It therefore made good sense to share the kill throughout the clan. Sharing reduced the risk of going hungry (because other hunters would reciprocate) at little cost, since there was plenty for everyone.

Cooperation, specialization and trade require friendliness. In the Stone Age, good guys finished first. Social skills, a propensity to trade information, to barter and do reciprocal favors: all these attributes are, to a greater or lesser degree, hardwired in us.

The friendly modern organization, which shares its largesse throughout the employee group, may therefore be going with the grain of human nature. It's good that we're programmed to cooperate, to be loyal and committed to each other, to be friendly to customers and collaborators; although a large organization with many different types of employees, may end up paying the less skilled more than it needs to. Worse, we find it difficult to deliver bad news, to measure each individual's contribution accurately, or to remove passengers. Perhaps we are 'wired' to favor equal treatment and 'fair shares for all' above efficiency and meritocracy.

On the whole, however, Stone Age friendliness and cooperation are highly appropriate to modern business. The same cannot be said of our other primitive characteristics.

Stereotyping on first impressions

Evolutionary psychologists say that because the Stone Age world was threatening and complex, it was necessary to classify things immediately on the most basic data. Which berries could be eaten or would poison you? Which regions were good for hunting? Which strangers could you trust? How do you decide? The only basis could be stereotypes based on first impressions. If people looked OK and acted friendly, they could probably be trusted. If not, they were probably enemies.

Anthropologists have found non-literate tribes that have classified every plant and animal in a similar way. This supports the view that successful Stone Age genes contained superior ability to make quick and mainly accurate decisions on first impressions. Sitting around to analyze the data was not life enhancing.

Today, it is not so vital to decide instantly. Whether Jill is a good person to hire does not have to be decided in the first 15 seconds of her initial interview. If we've scheduled an hour-long meeting to discuss a potential joint venture, we don't have to make our decision in the first minute. Yet research shows that we give enormous weight to first impressions. In turn, our immediate reactions affect the confidence and empathy returned by the people we meet, which reinforce our first impressions; the cycle is only likely to be broken if the other person says something with which we strongly disagree. As good salespeople know, a winning smile, a firm (but not too firm) handshake, and a good opening line can be more important than the intrinsic characteristics of what is for sale.

It is highly probable that we make many poor decisions, or fail to weigh the evidence judiciously, because of our Stone Age programming. We also waste the time we have. If we're going to go on first impressions, we might as well keep all meetings down to five minutes long.

Breast beating and mindless optimism

When life is random and terrifying, the person who appears least terrified and most confident is likely to attract followers, food and sex. Genes for confidence are likely to proliferate and be reinforced. Confidence becomes more highly valued and widespread than realism.

In the Stone Age, therefore, breast beating was not just an indulgence; it led to success.

The business world also responds to confidence. Yet although a mix of self-confidence, skill and luck can take someone a very long way, outside events often withdraw their cooperation. British Prime Minister Harold Macmillan, in the period when he apparently walked on water and was dubbed Supermac (after his contemporary, Superman), was asked what he feared. He was wise enough to reply, 'Events, dear boy, events.' And so it proved, in the form of his minister John Profumo, who had an affair with a prostitute and lied about it to Macmillan and (horror of horrors) to the House of Commons. How often has this pattern been repeated in business, with the most confident and successful leaders suddenly dragged down by events, especially when success has made the leader and the people around him or her blind to reality?

Yet to temper confidence with realism may be offensive to our genes. In terms of results, blind confidence ain't what it used to be. Nor are egotism and breast beating. They used to impress, now they merely annoy. Our environment has changed; we haven't.

A taste for hierarchy

Everything known about hunter-gatherer societies suggests that *ad hoc* hierarchies flourished. The desire to follow confident leaders and to find security in a chain of status relationships seem to be pronounced traits of primitive societies. Wealth was fluid and not easily accumulated, but food, shelter and sex gravitated to the leaders. For those who were not the leaders, the chances of security and an adequate living increased with attachment or deference to a leader. Herbert Simon, an economist and cybernetics expert, claims that 'docility' in early humans was the route to survival; by docility he means receptivity to social influence and the claims of leaders.

If we are wired for hierarchy, this helps to explain why every revolutionary attempt to dispense with hierarchy—whether the French or Russian Revolutions, or the modern flat, single-status organization—ends up replicating new forms of hierarchy. The twentieth century was both the 'century of the common man' and the century of the psychopathic leader, with more than 100 million people killed by the top three psychopaths (Stalin, Hitler and Mao) alone. Even in liberal democratic society and humanistic organizations, if official hierarchy is abolished, unofficial pecking orders spring up and flourish. Status is both sought and acknowledged.

The quote at the start of Robert Townsend's 1970 satire *Up The Organization*[5] reminds us:

And God created the Organization,
and gave It dominion over man.
—*Genesis 1, 30A, Subparagraph VIII*

The enduring taste for hierarchy may appear to be an anachronism in the age of the 'knowledge worker.' Evolutionary psychology helps to explain why hierarchy is devilishly difficult to extirpate.

Of course, the extent to which hierarchy fosters or hinders business success is very much a matter of opinion. I discuss this issue in Chapter 9, and the view I take is that hierarchy is sometimes extremely useful, when the boss has some genuine insight about business that is not generally shared. Hierarchy without insight, on the other hand, subtracts value. Most successful organizations manage to combine dictatorship of ends with democracy of means. Evolutionary psychology suggests that the latter is more difficult to engineer than the former.

Conformism and herding

Allied to hierarchy is the tendency for the clan to conform internally and be suspicious of those outside. High-status individuals are imitated. There is a natural tendency to do what the hierarchy wants. A sense of identification within the clan leads to cohesive behavior in the face of external threat. The individuals who thrived within any society were either leaders or docile followers; and the societies that thrived were those with the greatest internal cohesion. Conformity paid.

Conformity is still in the ascendant. Apart from the hundreds of millions of people who followed the twentieth-century psychopaths, at least partly because everyone else was doing so, we have only to look at the strength of successive conformist ideologies—socialism, imperial jingoism, antisemitism, McCarthyism, fundamentalist religions of all kinds—and at less harmful but equally absurd fashions such as kipper ties, flared jeans, the worship of rock stars or football teams, political correctness, or the excesses of both bull and bear stock markets, to realize that following the herd is as popular as ever.

This is familiar in organizations too. It takes a brave, foolish or unusually obstinate person to go against the received wisdom in any unit or firm. Remember William Whyte's 1956 business bestseller *The Organisation Man*? And from 1970, Robert Townsend's third paragraph jolts the reader:

> *In the average company the boys in the mailroom, the president, the vice-president, and the girls in the steno pool have three things in common: they are docile, they are bored, and they are dull. Trapped in the pigeon holes of organization charts, they've been made slaves to the rules of private and public hierarchies that run mindlessly on and on because nobody can change them.*[6]

Thirty years later, and after billions of dollars spent on gurus, consultants, and change programs, not much has altered. The management magazine *The Antidote* commented in 1999:

> *The organisation built on the industrial model was sabotaging its own efforts to get more initiative … out of its people. Despite an ever-expanding toolkit, best practices and what-have-you, research found that managers … were disillusioned with these attempts at fine-tuning yesterday's model … most of the core beliefs associated with the industrial model and 'organisation man' proved highly resistant to change.*[7]

Richard Pascale argues persuasively that a key weakness of most large organizations is their inability to tolerate and harness questioning and conflict. In his book *Managing on the Edge: How the Smartest Companies Use Conflict to Stay Ahead*,[8] Pascale estimates that half the time that contention arises, its potential value is lost because the conflict is smoothed over and avoided. Organizations are thus able to ignore important aspects of reality.

Hostility to out-groups

The strength of the clan and its conformity had the flipside of hostility to those outside. This, too, assisted survival, as Charles Darwin noted:

A tribe including many members who, from possessing in high degree the spirit of patriotism, fidelity, obedience, courage and sympathy, were always ready to aid one another, and to sacrifice themselves for the common good, would be victorious over other tribes; and this would be natural selection.[9]

Matt Ridley's study of primitive man leads him to a similar conclusion:

It is a rule of evolution … that the more co-operative societies are, the more violent the battles between them. We [humans] may be among the most collaborative social creatures on the planet, but we are also the most belligerent.[10]

Experience suggests that only trade and the weakening of exclusive national identities can lead to peace. Around the world war continues, despite the objective truth that conquering territory is no longer a sensible or necessary way to get rich. Our desire for group identity and bloody conflict with rival groups both persist, especially among young people, and are sublimated into demonstrations, football team rivalry, and, at another level, into business competition.

Yet even within business, what are we to make of the demonization of certain competitors, or the preposterously popular military analogies? John Kay points out that we talk about the 'cola wars' and yet that 'not in Pepsi's wildest fantasies does it imagine that the conflict will end in the second burning of Atlanta [Coca-Cola's head office].'[11] Business competition, happily, is not at all like war. In war, the biggest combatant usually wins, by destroying the opposition. In business, two or more competitors can flourish, and it is much more effective to avoid direct competition than to try to inflict direct damage on the competitor, a strategy that is rarely used and sometimes counterproductive even then. In business there is a third party, the customer, who decides which competitor will win: there is no such powerful third force in war. The idea that business is like war is so stupid that we can only explain the appeal of the analogy, like the continued appeal of war itself, by reference to our descent from Stone Age man, when belligerent, macho and destructive behavior paid off.

Neolithic habits die hard. Cohesion within a function, division or team is easy to build, but how often does manufacturing have a close relationship with marketing, or vice versa? The persistence of ancient rivalries should make us wonder whether an organization composed of several cohesive groups may not always be less effective, other things being equal, than one comprising a single homogeneous group.

The avoidance of risk

Evolutionary psychologists say that hunter-gatherers tended to take risks only when their world was falling apart. Their thesis is that our origins explain why today we are risk averse when we can afford risk, and yet risk takers when losses are endemic.

A high degree of security is necessary before we will take risks. This can be seen in the behavior of infants. Experiments in child psychology show that toddlers dare to explore only when their mothers are around.

For hunter-gatherers, these conditions of security rarely obtained. As long as they had about enough food and shelter, they wouldn't go hunting again until absolutely necessary. The risk of losing everything was just too great. Being eaten by a lion or attacked by one's prey can't have been an attractive risk to court deliberately. Because hunter-gatherers weren't secure, they generally avoided risk.

Now we are often much more secure than the hunter-gatherers, yet we are still loath to take risks, even when these are far from life threatening. A decision to leave a good corporate home, or to take a very different job within an existing firm, or to seek out different sorts of customers, or even to challenge the boss, may not in fact be very risky. Other ways of making our living are usually possible. Yet we perceive the risks to be greater than they really are. Quite possibly, this reluctance to take risks is a hangover from the Stone Age.

Risk aversion is built into most modern business. We talk about a 'risk premium,' where returns must be significantly higher to justify taking an extra risk. The concept is curious, given that the most important investors can afford to take risks, and can diversify their portfolios or hedge to keep risk within acceptable bounds; therefore, within certain limits, they should only care about the rate of return, not the risk-adjusted return. And if risk can be diversified away by holding a portfolio of investments, why should above-average volatility require a 'risk premium'?

There is a similar paradox in relation to equities and bonds (fixed-interest instruments). In the twentieth century, equities gave about a 6 percent higher rate of annual return than bonds, because equities were thought to be riskier. Yet the actual 'risk' of equities not matching the return on bonds, for any 10-year period, was negligible. In fact, the risk of getting an inferior return was nearly always greater for bonds than for equities. Only an atavistic aversion to apparent risk can explain the paradox. It is difficult for us to appreciate that uncertainty is not necessarily 'risky' in any real sense. Uncertainty can actually be very reliable: few businesses have such high and consistent returns as casinos. Placing a large number of bets, which is what casino owners do, need not be very risky.

There is a false market in risk. Because we are risk averse, risk premiums are higher than they should be. Individual entrepreneurs take risks, it is true, but generally only when convinced that the upside greatly exceeds the downside in both quantum and probability. Organizations and those in them are notoriously reluctant to take real risks, despite evidence that, increasingly, business fortune goes to the brave.

In high-growth markets, especially those involving networks, you have to place big bets to stand a good chance of winning. Up-front investment can be very high, because fixed costs are very high. Risk is also high, because each business segment will only support one substantial winner. But, equally, incremental costs are generally low (and sometimes virtually zero; on occasions, actually negative). The rewards from success can be astronomical, and quite disproportionate to the cost, even to the total cost of all the contenders, winners and loser. There is usually a first-mover advantage, and the first player to establish a significant lead (not necessarily the first mover) will usually be able to sustain its lead and go even further ahead. For an individual player, therefore, the risk of a low investment can actually be much higher than the risk of a high investment. The sensible choice is often between making a large bet, and making no bet at all.

Writer Thomas A Stewart, editor of *Fortune* magazine, advises firms not to try to avoid risk. He says that managing knowledge products is like

holding a book of innovative bets. Companies should maintain a portfolio of ideas in which risk is both maximized and diversified. Don't put all
your eggs in one basket, but do bet heavily.[12]

Mad scrambling when under serious threat

Primitive man did take risks and scramble furiously when his life was at
stake. Panic often worked. The more successful 'scramblers' survived more
often, so that scrambling under threat is part of our genetic make-up.

Today, such scrambling behavior may not be so functional. When non-
specific layoffs are announced, so that no one knows whether they'll keep
their job, productivity often rises sharply. When a factory closes, certain
employees may become hysterical or aggressive. When a normally sane
person is under pressure while driving, 'road rage' or sudden panic may
take over. Higher productivity, screaming at managers, and attacking a fellow motorist are all examples of mad scrambling when threatened; but
certainly the last two behaviors, and possibly the first, don't help the
scrambler at all. The modern world is not the Savannah Plain.

Another interesting example is gambling. Those who win modestly in
a casino usually want to cash in their chips before long. Gamblers who
start to lose often go beyond their preset limit of losses and gamble
everything they have available, in a mad scramble to get back to where
they started.

On the stock market, individual investors often display similar behavior. When their shares rise, they want to lock in the profit. If the shares fall
heavily, novices will nearly always hang on, in an effort to avoid serious
loss; they may even scramble to get funds enabling them to 'average down'
by buying more of the same shares at a lower price. Yet professional
investors are trained to take the opposite approach. It normally pays to run
your gains and cut your losses, not to fight the market. Trading houses and
fund managers often have strict rules to enforce such behavior; they need
to, because human nature tends in the opposite direction. When we are
losing badly, we are quite likely to gamble wildly. This is why Barings
Bank was brought to its knees by rogue trader Nick Leeson.

Further insight into our irrational attitudes to what we own, what we risk, and what we deserve are available from conventional psychology, biology, and game theory. I've picked three examples that are important for business and that illuminate, and to a degree corroborate, some findings from evolutionary psychology.

Owners and intruders

One fascinating insight from both biology and game theory relates to contests between owners or incumbents of territory, and challengers or intruders. Being the owner confers a psychological advantage that is independent of relative strength.

If two similar butterflies contest a sunny spot, they spiral around for a short time and then, typically, the intruder departs. But if both butterflies are already resident in the spot, the contest takes quite some time, with an equal chance of either winning. Experiments with baboons show the same pattern: the owner is likely to win, even if it is the slightly weaker animal. The reason is a strong inclination to hang on to what one already has, and a weaker inclination to take new ground.[13]

The endowment effect

Similar experiments on humans have shown what psychologists call the endowment effect. Say that someone has been offered two tickets to a top sporting event. A few days later she is offered $200 for the tickets. In experiments, most refuse the exchange. Put the experiment in reverse: give the subject $200, and then offer the same tickets in exchange for the money. What do you think happens? Apparently perversely, most still refuse the trade. But the idea of ownership explains the apparent paradox. What people are given first, and hence what they own, is what they want to keep.

Game theory (see Chapter 5) has validated the ownership advantage. In games of owners and intruders, there is a dominant strategy that in the long run will win most games. If you are the owner, you should escalate to deter the intruder. If you are the intruder, you should not escalate. In fact, if the owner escalates, you should retreat. An owner should make a stand, because the intruder is then likely to retreat. When players swap roles, the same phenomenon occurs: the owner (previously the retreating intruder) tends to escalate and win; and the new intruder (previously the fearless owner) retreats. Eminent biologist John Maynard Smith called this pattern 'bourgeois' competition, since respect for property rights ensures that conflict is not escalated. Bourgeois competition is implicit cooperation, with consent given to incumbents.

Clearly, this is useful knowledge in business. There is an inbuilt, apparently hardwired, genetic bias towards respect for incumbents that places challengers at a disadvantage. This bias is probably related to the risk aversion noted above: when hunter-gatherers had enough they would not fight for more, because it was too risky; but they would scramble to defend what they already had, because this was vital for their security. Now there is no objective basis for our behavior, but it still reflects Stone Age conditions. Consider these corollaries:

◆ The so-called 'first-mover' advantage—the well-documented tendency for the first firm into a market to enjoy an advantage usually greater than it deserves on objective merit alone, independent of being first—may be rooted as much in the psychology of the challenger and the incumbent as it is in the effect on customers. The conventional explanation for first-mover advantage is that customers associate the new niche with the first firm into it. Obviously, not much can be done about this by the challenger. If, however, much of the advantage resides purely in the mind game between challenger and incumbent, the advantage may not be as secure as it seems. If the incumbent can somehow psych itself up to an absolute determination to take the territory, and make it clear to the incumbent that it is no respecter of 'bourgeois' rights, then the odds may shift significantly in its favor.

◆ Under current behavior, however, if the incumbent and challenger are otherwise equally matched, the incumbent enjoys a clear advantage, if only in terms of motivation. If the incumbent comes under attack, even from a superior product or service, it should escalate the conflict, knowing that the odds are that the challenger will back down.

◆ Perceived ownership is what matters. It is not relevant which butterfly is the real owner of the sunny glade. The butterfly that considers it belongs there will tend to win over the butterfly that thinks of itself as an intruder. If you want your firm to win in a particular market, therefore, you have to drill it into your people that you belong there, that it is rightfully yours. Bending the facts, as long as you can do so convincingly, can be very useful; propaganda can decide the contest. People will not easily concede space that they believe is theirs. On the other hand, they will not put their best foot forward if they think they are trespassing.

The ultimatum bargaining game

Robert Frank, an economist interested in explaining why people do things using emotional as well as rational explanations, draws attention to the issue of fairness and the role it plays in negotiations. Psychologists use an experiment called the *Ultimatum bargaining game*.

This is a game for two people, say Alex and Justin. Alex is given $500 and told to share it with Justin. If Justin accepts what he's given, they both keep the cash. If Justin refuses the deal, neither gets anything. What should Alex offer Justin?

If Alex and Justin are clinically 'rational,' and do not allow emotion to creep in, Alex should offer Justin a very small cut, maybe $5, and keep the other $495. After all, Alex has the cash and Justin has the choice: $5 or nothing. Dr. Spock would certainly advise Justin to accept.

The thing is, in experiments where this happens, Justin tends to refuse the $5, and both players go away empty handed. Apparently, our deep-rooted sense of fairness is offended by the inequality of the split, so that Justin looks at the *proportion* he's getting, rather than the absolute amount.

Not only that. In practice, Alex doesn't go for the rational solution and offer a small share to Justin. By far the most common outcome in experiments is that the person with the money offers $250, half the cash.

Does this behavior reflect the sharing of meat on the Savannah Plain, where the hunters are thought to have shared with those who didn't participate in the kill? Are generosity *and* socialistic envy hardwired by natural selection, where evolution has not had time to catch up with modern conditions?

At work, do we expect to have to share even where our colleagues have made no contribution? And do the latter get indignant if they don't get a 'fair' chunk of the bonus pool, even if they've personally had a lean year? I've observed this behavior many times when advising bosses who have to decide bonuses. Justin may have brought no profit this year, but we don't want to demotivate him by a derisory bonus. Such thinking is deep rooted and frustrates the very purpose of meritocratic rewards.

I believe that large organizations are inevitably socialistic, in that they find it impossible to reward according to just deserts. Instead, big business builds in an undue share of the spoils for the passengers and for those who are not vital. In large multinational corporations, non-core workers, including those engaged in tasks such as cleaning, routine maintenance or clerical functions, are generally paid substantially more than they would be for the same job outside the multinational, because otherwise the gap between them and their more skilled colleagues would seem too large. The dilemma faced by those who set pay is this: do we reward according to our economic needs and the laws of supply and demand, or do we do what people (wrongly) regard as fair? Large organizations almost always go down the second route. Indeed, one of the reasons that 'outsourcing' is so popular is that it offers a neat way out of the dilemma. If the cleaners and clerks and maintenance workers are outside the organization, we only have to pay them their market rate, however high the compensation of the knowledge workers inside.

Are we really hardwired?

There are two interesting contemporary scientific challenges to the 'hardwiring' theory. One is the increasing evidence that learned behavior can win out against instinct; that the dominance of our genes is being undermined. The other challenge is from the emergent science of neuroplasticity, which suggests that, far from being hardwired, the brain can be *re*wired by the mind.

The theory of cultural evolution in animals

We came across the first of these challenges in Chapter 2. Biologists such as Richard Dawkins and E O Wilson have suggested that in human society, learned behavior ('culture') can operate independently of genes. Now other biologists are going much further, and arguing that many other animals, including some very unsophisticated organisms, may exhibit 'cultural evolution,' which often runs directly counter to genetic preferences. Their experimental evidence is fascinating.

A Canadian marine biologist, Hal Whitehead, suggests that sperm whales learn songs from their mothers.[14] As we saw in Chapter 2, biologist Lee Alan Dugatkin has shown that female guppies tend to mate with the male guppies that have already been selected as mates by other females. The females will even switch preferred mates in order to copy other females. Dugatkin shows that learned behavior has invaded the province of genes in marine bugs and many types of birds.[15] His conclusion is:

> *Cultural norms may take on a life of their own and 'run away' in very unexpected directions—directions that were never even in the picture for our primitive ancestors roaming the Savannah of Africa.*

Neuroplasticity

Of even greater interest is **neuroplasticity**, which claims that the brain can rewire itself.

Children have an innate ability to learn language. But which language they learn, claims Dr. Jeffrey Schwartz,[16] determines how the brain stores sound. Now neuroscientists understand why: the particular sounds made by the language in question vary (Japanese sounds are very different from English ones), and each language forces a different rewiring of the part of the brain that processes language.

Brain scans also show that obsessive–compulsive disorders can be moderated or cured by deliberate action by patients to think about other things. The brain circuits can actually be rewired by conscious effort. The brain is therefore plastic. Jeffrey Schwartz also claims:

> There is a mind independent of the brain … if the mind can rewire the brain, then in an important sense the mind is master of the brain.

Neuroplasticity is even more controversial than evolutionary psychology; the concept of rewiring cannot be said to have replaced that of hard-wiring. But how do the theories of cultural evolution and neuroplasticity affect what we can take away from evolutionary psychology? And may they have their own lessons for business?

Transcending genetic dispositions

I don't see any necessary contradiction between the implications of evolutionary psychology, on the one hand, and those of cultural evolution or neuroplasticity on the other hand. In fact, from a business perspective, I find the three disciplines complementary. The explanations for human behavior provided by evolutionary psychology resonate with business experience, especially that of men within corporations. This may be entirely a coincidence; but more likely, I think, it is because evolutionary psychology is at least partly right. There probably is a genetic predisposition towards behavior that made more sense on the Savannah Plain than it does today.

Nevertheless, this view is not incompatible with cultural evolution. Humans can *learn and transmit learning*; this means that genes are important in determining our behavior, but not that all important. If we have not yet learned to control and correct for all of our dysfunctional genetic inclinations, this does not mean that we cannot in the future, just as we have learned to welcome strangers on first sight. If we understand what we are up against, in terms of genetic bias, we are actually much more likely to be successful in correcting this bias.

Finally, the conflict between hardwiring and rewiring may be more apparent than real. Why can't we be both hardwired *and* capable of rewiring at least part of this hardwiring? Clearly, the brain is not totally plastic—but why should it be totally hardwired? And even if we really are totally hardwired, it may be functional to believe the opposite. We can change our behavior even if we cannot change our genes or our brains. I am confident that it could be empirically proved that fatalistic business-people are less successful than those who believe in free will.

I predict that executives will hear a lot more about cultural evolution and neuroplasticity. Part of the argument of this book is that we are moving beyond mechanical, cause-and-effect models—which are, and will remain, valuable—to embrace in addition more fluid, biological models. But beyond the biological models, there probably lie neurological models, which are even more free-form and sensitive to creative manipulation.

How to manage and mutate Stone Age man

My conclusion, therefore, is that we *can* manage and mutate Stone Age man, and that it is well worth the effort to do so. If we accept the insights from evolutionary psychology, but also the possibility of change, what should we do differently? Some suggestions:

◆ We should recognize ourselves and our colleagues as what we are: twenty-first-century impostors driven by neolithic genes. The genes are sometimes helpful to business, as when they facilitate cooperation, but are generally a nuisance: they obstruct commercial progress and the good of our firms and our careers.

◆ We should start with ourselves. We have met the Flintstones and they are us. We should make continual and vigorous attempts to correct for our dysfunctional tendencies, like the undue avoidance of risk, the rejection of criticism, and the tendency to jump to conclusions about people on the basis of first impressions.

◆ If we wish to influence other people, we cannot appeal only to their reason. We must be skilled at baser appeals to their emotions. If we wish to manage people, we must manipulate their primitive propensities.

There are a number of structural remedies that help simulate the Stone Age clan, based on the general proposition that, all other things being equal, large, complex, and disparate organizations will never work as well economically as smaller, simpler, and more homogeneous ones.

◆ First, we should accept that the natural capacity of a cohesive unit is up to about 150 people—the size of the largest hunter-gatherer clans. When a unit holds fewer than 150 people, everyone can know everyone on first-name terms. Percy Barnevik of the Swiss-Swedish engineering concern Asea Brown Boveri (ABB) has achieved great success by breaking up his massive empire into units of 50 people. Microsoft goes even further, with a proliferation of small units, each having typically only five to ten people. We tend to forget that, despite all the advantages of large firms with global reach, 60 percent of all employees worldwide work in small to mid-size businesses, often owned or run by a family.

◆ Second, firms above the ideal clan size tend to divide into separate functional, regional or product 'clans,' which frequently squabble with each other. The near-universal experience of managers is that it is more difficult to do business with sister clans in the same organization than with strangers freely chosen by each clan.

◆ Third, organizational size, complexity, and heterogeneity lead to inefficiency *because of the expectations and behavior of the people involved.*

A large organization contains more people than a small one, and it is generally more difficult to detect or measure the real economic contribution of each person. This technical difficulty is exacerbated because of the socialistic tendency of organizations to meet 'fair' expectations of employees based on a claim for a share of the pie simply because of clan membership, regardless of contribution.

The more complex an organization—the more things it does, the more customers and suppliers and products it has—the more difficult it becomes to measure and reward true individual contributions. A complex organization of a thousand people will have far more difficulty than a simple one with the same number.

Then consider heterogeneity, which is not the same as complexity. Complexity relates to what the organization does, heterogeneity to what the organization is, in terms of the employees' composition. A heterogeneous organization is one comprising many different types of people, with different backgrounds, disciplines, degrees of skill level—for example, an organization with 10 main disciplines such as engineering, computing, marketing and so on, but comprising exclusively graduates, may be more homogeneous than another organization with only three disciplines but three different sets of educational and class backgrounds, nationalities, and styles. Evolutionary psychology suggests that people don't easily like or trust those from different camps.

These three forces of division and non-economic claims on resources—size, complexity and heterogeneity—compound each other. If a firm is very large, *and* very complex, *and* very heterogeneous, it will have to deal with a terrifying amount of internal conflict, which can only be suppressed or appeased at high cost. It won't be able to properly reward its real benefactors among all its employees, customers, suppliers, and other collaborators. Simple, focused, and homogeneous firms are much better placed to give credit where it's due.

Incidentally, it is interesting that small national size carries no economic disadvantage, perhaps because scale effects are less important than simplicity and common identity. Of the 10 wealthiest (per capita) nations, apart from the United States, the largest is Belgium, which has only 10 million people. Of the 10 countries with more than 100 million people, only the United States and Japan are prosperous.[17]

Remember the lesson from Chapter 1, that diversity works, and that from Chapter 2, that the most successful organizations have a varied gene pool. How do we synthesize these insights with those from evolutionary psychology? I think we have to conclude, whether we are thinking of organizations or of societies, that heterogeneity is enormously valuable, but that it needs to be *managed*. Our genes are not cooperative or functional in this respect, so we need to interpose management mechanisms, culture, and vigilance to keep reminding ourselves of the value of variety. I think there is a second point too. Heterogeneity goes against our genes, yet is valuable. But the size and complexity of many organizations also go against our genes, and have no compensating benefits. The best solution is to have small, simple, and heterogeneous organizations.

So yes, we can manage and mutate Stone Age man. We manage by manipulating and correcting for our primitive instincts. We mutate by changing the business context, to capitalize on the good parts of our genetic heritage and to minimize the damage from the harmful parts. Only by recognizing what we are up against can we raise our managerial game.

Summary

Evolutionary psychology lights up the lunacies of business life. Forget the organizational charts and the idea that corporation is a rational economic entity. Gaze instead on a seething cauldron of over-sized egos preying on submissive souls, where appearance and confidence can crowd out substance and contribution, where conformity to the clan and hostility to other clans undermine rational decisions, and where risk is routinely avoided until panic sets in. Evolutionary psychology explains why large organizations are so difficult to change, and why we cannot be relied on to act in our own best interests.

Life was much harder on the Savannah Plain. But it was also much simpler, and at least the genes worked in the right way. Yet fear not. Cultural evolution tells us that we develop and imitate more successful economic formulae, starting with firms that are small and simple. Neuroplasticity says that we can rewire our own brains by deliberate acts of will. Our neolithic genes are not our best friends, but they need not be our masters. The sophisticates' revolt is at hand, fueled equally by scientific reason and self-friendly willpower.

Action implications

◆ *Manage and mutate Stone Age man*, starting with yourself. Listen to criticism. You have no need to defend yourself.

   Allow second and third impressions to overrule first impressions.

Take more risks. If the upside—value times probability—exceeds the downside, do it. If in doubt, do it.

The one exception is not to take risks when things are going badly. Don't panic. Just cut your losses and get out.

Don't be a prisoner to your emotions. Don't take decisions in the heat of the moment. Take time to calculate what is best for you.

Don't thump your chest. Don't draw attention to your perspicacity or achievements. Don't tell people that you were right.

When you are successful, beware. Don't believe that you are infallible. Remember how important luck is. Expect reverses and take them in your stride.

◆ *Capitalize on the genetic predispositions of others.*

Act confidently, even when you are not confident. But don't believe your own propaganda.

If you have an insight and want to lead, even among equals or superiors, go ahead and lead. Most people like to be led. The informal hierarchy is more powerful than the formal one.

Be careful to present yourself so that others' first impressions correspond to what you want. If you want them to think you dynamic, move fast. If you want to appear scholarly and reflective, wear tweed coats and glasses. Wear your façade with pride.

Be friendly and warm, especially when first meeting someone. Try to read their mind and empathize.

Be willing to gossip. Share information. Build trust. Don't criticize other people to third parties.

Try to avoid direct criticism of someone. Always praise first, to make them receptive. Criticize actions, not the person.

Only openly disagree with those below you in the pecking order. Otherwise, use the venerable line, 'I agree with you, but…'

Build personal alliances throughout the organization, especially in unlikely places. If you are an engineer, hang around marketing until they accept you as one of them.

Realize that reputation is more important than performance. Reputation is three parts identification/empathy and one part competence.

◆ *Manage Stone Age folk appropriately—expect and correct for irrationality.*
Be aware that the hierarchy may induce conformity and blindness to reality. Use formal hierarchy sparingly, and encourage dissent. Don't promote authoritarian characters to positions where awareness of the outside world is important.

When hierarchy is subverted by an unofficial pecking order, don't try to resist.

Don't seek to eliminate all forms of hierarchy. Don't expect delayering and single-status workplaces to usher in a new age of egalitarianism and democracy, or 'empowerment' to generate initiative. Allow the natural rhythm of leadership and followership to operate. Allow people to work together naturally, selecting their own *ad hoc* alliances and teams.

Be profligate and universal with status awards. Status is like flattery: no one is too smart or savvy to avoid its charm. Give special responsibilities, such as leader of project team X. Identify and celebrate unusual achievements, such as 'beyond the call of duty' service to customers.

Encourage the natural tendency towards internal cohesiveness. Talk about colleagues and not employees. Encourage colleagues to say 'we' not 'I,' 'us' not 'you.'

Anticipate hostility towards out-groups. Focus it on competitors rather than other departments or functions within the firm. Sometimes hostility is focused on another out-group, the customers, with obvious consequences. Reserve your most severe sanctions for discouragement of this behavior. Encourage colleagues to think about customers as an extended part of the clan, as part of the club, along the lines of quality engineer W Edwards Deming's dictum: 'The customer is the most important part of the production line.' Encourage all employees to see as much of customers as possible, because familiarity breeds affection.

Counter the out-group feelings between different parts of the organization by forming cross-group project teams to achieve worthwhile ends. Make the teams as heterogeneous as possible.

Rotate people's jobs internally, with the explicit aim of putting different types of people into each clan (e.g. marketing people into manufacturing).

When hiring, ensure that everyone guards against first impressions and the clone mentality. Don't allow anyone to give their impressions of candidates to someone who has not yet interviewed them. Use objective testing and outside assessment.

Question any internal consensus. It's probably wrong.

Expect but discourage macho behavior: breast beating, one-upmanship, self-aggrandizement, empire building, malicious gossip, personalized competition. The best way to counter macho behavior is to hire as many women as possible, so long as they have not succeeded by becoming like men.

Don't expect people to take risks or welcome change or be creative. If risk taking is essential, frame the situation as an emergency. Make it life threatening (or at least job threatening). Be up-front and even exaggerate any bad news: 'Competitor *X*, if left unchecked, will have taken half our business within two years.' 'If return on capital doesn't double, we'll have to downsize.' 'This new technology could kill us.' 'Customer *Y* accounts for 60 percent of our profits but is very unhappy and ready to take its business away.' 'If we don't internationalize, we'll be taken over.' Choose whatever is credible, really worrying, and at least half true!

Make it clear that failure through taking risks is more acceptable than refusal to take risks.

If you want creativity, make the environment non-threatening and informal. (Children are creative, but only so long as they feel totally protected.)

◆ *Change structures in order to counter primitive genes.*

Reduce your firm's size and complexity.

If different parts of the firm have different ways of working and core competencies, split it into two or more new firms.

Ensure that each organizational unit within the firm has no more than 150 members.

Make the organization structure as simple as possible. Realize that matrix organizations come up against primitive man, and the latter wins. People gravitate towards one main sense of identity and loyalty. Therefore, use matrix forms sparingly and realize that there will always be one main affiliation and one less important dotted-line one.

When risk taking is important, form a separate unit that has to take risks. Make it clear that membership of the unit is temporary and that there is a way back to the parent whether the fledgling business succeeds or fails.

♦ *Use the specter of competition to rally your troops—but avoid fighting.*

Realize that business is not war; or that if it is, that fact is purely for internal consumption. In George Orwell's novel *1984*, there are three superpowers and two of the three are always at war with the third. From time to time the sides switch. There is no evidence of actual conflict, yet there are perpetual reports of victories and calls for redoubling of home efforts to support the war. This is not a bad model for business.

Collaborate with competitors as far as possible. Avoid direct conflict or damage.

Don't put your competitor's back against the wall. This may precipitate mad scrambling, which may or may not help the competitor, but will certainly damage your firm.

Remember that the only useful way of 'beating' competitors is by having your own loyal fan club of profitable customers, who prefer what you do to anything any other competitor does.

Notes

1 The real problem, evolutionary psychologists imply (although they are often coy about saying so bluntly), is not so much Stone Age woman as Stone Age man. There is an unfashionable sexism that is implicit in evolutionary psychology, because one of its contentions is that sexual roles too are hardwired and that we cannot totally escape them. The macho nature of Stone Age man is more inappropriate to today's conditions than the more passive and modest behavior of women. Objectively, therefore, women may be better suited to business than are men. But because men dominate organizations and set the rules, women may find it difficult to conform and break the glass ceiling. This may explain why there are plenty of examples of successful female entrepreneurs, yet few women at the helm of big business.

2 See Steven Pinker (1997) *How the Mind Works*, Norton, New York; Matt
   Ridley (1996) *The Origins of Virtue*, Viking/Penguin, New York/London; and
   Robert Wright (1994) *The Moral Animal*, Little, Brown, New York. See also
   the terrific article from which many of the points in this chapter are derived:
   Nigel Nicholson (1998) 'How hardwired is human behavior?' *Harvard
   Business Review*, July–August.

3 Robin Dunbar (1996) *Grooming, Gossip and the Evolution of Language*, Faber
   & Faber, London.

4 Matt Ridley, *The Origins of Virtue*, p 95.

5 Robert Townsend (1970) *Up The Organization*, Michael Joseph, London.

6 Ibid., p 9.

7 *The Antidote*, Issue 19, 1999, p 10. The article is reporting on the views of
   Sumantra Ghoshal and Christopher A Bartlett (1998) *The Individualized
   Corporation*, William Heinemann, London.

8 Richard Pascale (1990) *Managing on the Edge*, Simon & Schuster, New York.

9 Charles Darwin (1871) *The Descent of Man and Selection in Relation to Sex*,
   John Murray, London.

10 Matt Ridley, *The Origins of Virtue*, p 193.

11 John Kay (1999) 'Total war and managers from Mars,' *Financial Times*, August
   4.

12 Report on a Strategic Planning Society conference, *Don't try to minimise risk*,
   in *Strategy*, January 1999 (The Strategic Planning Society, London). See also
   Thomas A Stewart (1998) *Intellectual Capital: the New Wealth of Organizations*,
   Nicholas Brealey, London.

13 This example and the endowment effect are taken from a charming book:
   Karl Sigmund (1993) *Games of Life*, Oxford University Press, Oxford,
   Chapter 7.

14 Philip Cohen (1998) 'Song lines: singing lessons could affect the evolution
   of whales,' *New Scientist*, 5 December, p 15 (www.newscientist.com).

15 Lee Alan Dugatkin and Jean-Guy J. Godin (1992) 'Reversal of female mate
   choice by copying in the guppy,' *Proceedings of the Royal Society of London*,
   249:179–84. See also Dugatkin's forthcoming book *Guppy Love: Genes,
   Culture and the Science of Mate Choice*.

16 See the forthcoming book by Jeffrey Schwartz and Sharon Begley, *The
   Mindful Brain: a New Paradigm for Understanding How the Mind Rewires the*

Brain. See also Josie Glausiusz (1996) 'The chemistry of obsession,' *Discover*, June, p 36.

17 See 'Small but perfectly formed,' *The Economist*, January 3, 1998.

5

On Resolving the Prisoner's Dilemma

If men were actuated by self-interest, which they are not—except in the case of a few saints—the whole human race would co-operate. There would be no more wars, no more armies, no more bombs.

Bertrand Russell

Game theory

We turn now to our most uplifting subject: the search for cooperation, especially human cooperation. Here we supplement biology with **Game theory**, a branch of mathematics with links to many other scientific disciplines, including biology itself. The main lesson is how to cooperate effectively in pursuit of entirely selfish ends. We'll reach some pretty intriguing, counterintuitive, distinctly useful and pleasing conclusions, because game theory itself has a history, with sinister beginnings and a happy ending.

We can then relate game theory to our earlier observations on evolution and business genes, culminating in a theory of human evolution combining science, business, and human cooperation.

Game theory started when John von Neumann, a Hungarian genius who was also one of the architects of the computer, published his mathematical *Theory of Parlor Games* in 1928. Game theory is thus a branch of mathematics and statistics that has since been applied to economics, biology, epidemiology, philosophy, physics, politics, the social sciences, military strategy, and also business strategy.

Game theory deals with games where the most profitable thing for you to do depends on what your opponent does, and vice versa; it attempts to simplify the world and produce the best outcome, mathematically derived, for any particular situation. In 1944, von Neumann and economist Oskar Morgenstern published *The Theory of Games and Economic Behavior.* They invented the concept of the non-zero-sum game, where it pays to collaborate and form coalitions.

The Prisoner's Dilemma

The most famous 'game' in game theory, one that had been understood, in non-mathematical terms, for centuries—the basic idea surfaces in Hobbes, Rousseau and also Puccini's opera *Tosca*—is the **Prisoner's Dilemma.** There are many, many variants of this game, but the basic idea is the same.

Imagine two criminals caught and locked up in separate rooms. A robbery with murder has been committed, but the police have dubious evidence. If one of the criminals confesses before the other, the authorities offer a deal: immunity from prosecution versus execution for the other criminal. If neither rats on the other, they can both expect five years in prison for the robbery. The rational thing for each individual to do is to 'defect' as soon as possible (defect here means 'fail to cooperate' or 'rat'). Cooperation between the parties is irrational for any individual prisoner, although if they both cooperated neither would be executed.

Self-interest rules OK?

The game can be expressed mathematically, with rewards rather than punishments. Assume that Brian and Lee are playing and that you are

Brian. You are both invited to write down, simultaneously, either the number 1 or the number 2. If you write 1 and so does Lee, you both get $5. If you write 2 and Lee writes 1, you get $20 and Lee gets nothing. If you write 1 and Lee writes 2, the opposite occurs: you get nothing and Lee gets $20. Finally, if you both write 2, you each get $1.

What do you do? You'll probably write down 2, reasoning as follows. If Lee writes 1 and you write 1, then you will win $5; but if you write 2 (and Lee writes 1) you'll win $20, so 2 is better in this case. What happens if Lee writes 2? If you write 1, you'll get nothing, but if you write 2, at least you'll get $1. So whether Lee writes 1 or 2, you'll be better off if you write 2.

But the dilemma is that Lee is thinking the same way, since the payoffs are entirely symmetrical. If he follows self-interest, he'll also write 2. So you end up with $1. Yet if you had cooperated, you could each have had $5.

The conclusion of the Prisoner's Dilemma Mark I is that although mutual cooperation may be in everyone's aggregate interest, self-interest will tend to predominate, to society's disadvantage. The Prisoner's Dilemma was formalized as a game in 1950 by the RAND corporation in California. It was soon realized that the world was full of Prisoner's Dilemmas. Trees in tropical rainforests spend all their energy growing towards the sky rather than reproducing. If the trees could agree not to grow more than ten feet tall, each tree would still enjoy the same sunlight and they could divert the surplus effort into tree sex. But they don't.

The conclusion of the Prisoner's Dilemma, up to the late 1970s, was deeply depressing. The economics of the past 200 years had been built on self-interest. Yet now it appeared that self-interest was suboptimal. The dilemma was that it was inevitable. This appeared, too, to be confirmed by a lesson from evolution: the **Red queen effect**, which has also been called the 'evolutionary arms race.'

The red queen effect

In Lewis Carroll's *Through the Looking Glass*, the red queen has to run as fast as she can just to stay in the same place. This is similar to running up an escalator that's going down. It's also how evolution works in animals.

Lions chase antelopes. Lynxes chase rabbits. Over time, the antelopes and the rabbits get faster. Why? Because the antelopes and rabbits that are faster than their peers survive longer, and pass their genes on to the next generation. This is great for antelopes and rabbits, but it doesn't improve their position *vis-à-vis* predators. The same improvement in speed happens via natural selection for the lions and lynxes. So the 99th generation of rabbits flee faster from the 99th generation of lynxes, yet are in no less danger than their ancestors.

Richard Dawkins calls this the **Evolutionary arms race.** There is constant escalation and improvement in the arms of both sides of the predator–prey divide, but no change in relative position.

Unfair to antelopes and rabbits? Maybe. That is the price of progress via natural selection.

The evolutionary arms race applies to business too. Executives today work much harder and much longer hours than they did when I left university. There is no net individual advantage in this.

Executives and firms have to keep getting better just to stay where they are. If an organization bucks this trend—it manages to hold market share, despite not improving what it offers customers—this must mean that it's not in a competitive marketplace. How long will this last?

If you went through the business plans of the five largest competitors in any market, it's almost certain that all five would be planning to gain market share. Each company is going to be better than it used to be, so it'll get more business, won't it? This seems such a reasonable assumption, yet it is plainly impossible for everyone to gain share. The business plans fail to factor in the red queen effect. No one else is going to stand still. Your market share escalator is programmed to go down, and doing things better may only keep you where you are.

But is there really no escape from the evolutionary arms race? For animals other than humans, no. But for humans, and for business, it may be different.

The Prisoner's Dilemma revisited

The evolutionary arms race occurs for precisely the same reason that the Prisoner's Dilemma can lead each individual to act against the collective interest: the inability to cooperate. If lynxes could cooperate with rabbits, they could call a moratorium on getting faster, and devote their evolutionary efforts to some more beneficial objective. But of course, lynxes can't collaborate with rabbits to defeat the current version of evolution.

And yet, animals do collaborate for some purposes. In the 1970s, economist and geneticist John Maynard Smith turned to game theory to explain why animals typically don't fight to the death. His innovation was to play the Prisoner's Dilemma many times.

By playing a version of the Prisoner's Dilemma with hawks playing doves, and replaying the same many times, he showed that the best results were obtained by 'Retaliator,' a dove that turns into a hawk when dealing with hawks.

Initially, Maynard Smith was ignored. Then game theorists started using computers to play Prisoner's Dilemma games over and over. In the late 1970s, tournaments were organized with competing computer programs that played the game 200 times. To general surprise, the 'nicer' or more cooperative programs tended to win. The strategy that came out top was 'Tit-for-tat,' devised by Canadian political scientist Anatol Rappoport. Tit-for-tat was similar to Retaliator; it started by cooperating and then mimicked the last move of the other player. The organizer of the tournaments explained the success of Tit-for-tat:

> What accounts for Tit-for-tat's robust success is its combination of being nice, forgiving and clear. Its niceness prevents it from getting into unnecessary trouble. Its retaliation discourages the other side from persisting whenever defection is tried. Its forgiveness helps restore mutual co-operation. And its clarity makes it intelligible to the other player, thereby eliciting long-term co-operation.[1]

Long-term advantage often requires cooperating players to 'take turns' in collecting the payoff: I let you win the biggest prize this time, perhaps taking nothing myself, if you let me take the biggest prize next time. Cooperation is about comprehending how to make the pie bigger, on the understanding that when we have to divide it, we will behave reasonably, within the context of a long-term relationship.

Ridley's theory of social coagulation

In 1996 came one of the most important books of the decade: Matt Ridley's *The Origins of Virtue.*[2] This is a thesis on cooperation and virtue, drawing lessons from biology and economics. It also has profound implications for business.

Ridley gives a unique twist to the ideas of the selfish gene and evolutionary biology. He argues that society is a product of our genes and our evolution. Humans are unique because we are organized into large groups with complex inter-relationships between individuals—and because we cooperate in a qualitatively different way from all other animals. This is the highest and most successful form of evolution. As he says:

> *The essential virtuousness of human beings is proved not by parallels in the animal kingdom, but by the very lack of convincing parallels.*

The advantage of society rests in division of labor and socialization, which modern humans have taken to a happy extreme. Ridley again:

> *To thrive in a complex society, you need a big brain. To acquire a big brain, you need to live in a complex society. The human brain is not just better than that of other animals, it is different ... in a fascinating way: it ...[can] exploit reciprocity, to trade favours and to reap the benefits of social living.*

Division of labor

If you've studied any economics, you'll probably recall Adam Smith's example of the pin factory where 10 people, through specialization, were able to produce 48,000 pins a day. Smith was a Scottish political philosopher, and his 1776 book *The Wealth of Nations* invented the concept of division of labor:

> *The greatest improvement in the productive powers of labour, and the greater part of the skill, dexterity, and judgment with which it is any where directed, or applied, seem to have been the effects of the division of labour.*

Smith posits three key advantages of specialization:

♦ Practice leads to higher productivity.
♦ Specialization saves time switching from one task to another (which is, incidentally, the main insight behind business process reengineering, a major source of productivity improvement in the 1990s).
♦ Specialization makes it worthwhile to invest in tailored machinery to boost productivity, thus enabling 'one man to do the work of many.'

Smith demonstrates—and here comes the link with increasing degrees of cooperation—that:

♦ division of labor is limited by the size of the market, and can therefore increase when market size increases; and
♦ division of labor increases with better transport and communications.

It turns out that division of labor and increasing specialization and 'trade' are themes noted by biologists and anthropologists in explaining how complex organisms evolve and thrive. Increased cooperation is both a cause and a result of such success.

Biology reveals that the bigger the cells in small organisms, the more likely they are to divide the labor, with some cells specializing in reproduction. The social insects—like ants, termites, and bees—have been so successful because their societies contain specialized roles that require cooperation.

Human societies organized into bigger groups have more different jobs. The isolated and now extinct Tasmanians had only two castes and lived in groups of 15 or fewer; but the Maoris, who lived in groups of up to 2000, recognized 60 different professional roles.

Ricardo's law of comparative advantage

The links between cooperation, division of labor, and trade were first made explicit in 1817 by David Ricardo, a very rich investor, economist, and radical British politician. His law of comparative advantage applied division of labor to groups and countries. The law is impressive because it is counterintuitive—Paul Samuelson, a famous economist, says that Ricardo's law is the only proposition in the whole of social science that is both true and not trivial!

Until Ricardo, it seemed obvious that countries could only trade if one was better than the other at producing something. Ricardo said that in fact there was a basis for trade whenever the relative ratios of productivity were different, whatever the absolute levels—which implied that there was virtually no limit on possible constructive trade. If country X was better than country Y at producing two products, there could still be trade between them that would enrich both countries. If country X is twice as productive in steel and four times as productive in leather goods, then country X should specialize in leather goods and country Y should specialize in steel, where it had comparative advantage, despite being inferior in absolute terms.

Specialized business and trade—which are essentially cooperative inter-group activities—lie at the heart of human advances. As Matt Ridley explains so well:

> *200,000 years ago, stone age tools were travelling long distances from their quarries ... Where the Neanderthals all lived in much the same fashion, their replacements began to show great local variations in their stone technologies and styles of art ...*
>
> *[The] invention [of trade] represents one of the very few moments in evolution when* Homo sapiens *stumbled on some competitive ecological advantage over other species that was truly unique. There is simply no other animal that exploits the law of comparative advantage between groups. Within groups ... the division of labour is beautifully exploited by the ants, the mole rats, the Huia birds. But not between groups.*
>
> *David Ricardo explained a trick that our ancestors had invented*

many, many years before. The law of comparative advantage is one of the ecological aces that our species holds.

The importance of trade and cooperation can be graphically illustrated by contrasting the native Tasmanians to the native Aboriginal Australians. When Europeans landed on Tasmania in 1642, they ended 10,000 years of isolation, and found the most primitive human society in the world. Native Tasmanians couldn't light a fire from scratch. They didn't have bone tools or multipiece stone tools or axes with handles or boomerangs. They didn't know how to fish or make warm clothing. The Aboriginal Australians could do all these things; only Tasmania's isolation had kept its population poor and low.[3]

Isolated groups suffer, because most groups get their ideas and innovations from outside, via trade in ideas and goods. Human society progresses via increased trade, specialization, and inter-group cooperation.

Beyond baboonery

Ridley observes that two junior male baboons will join forces to beat off the consort of a female baboon. Then they chase after the female, and the one who catches her has sex. Cooperation is used to achieve selfish ends. This is the basis of our cooperative instincts.

Ridley takes up the theme expressed by Richard Dawkins in his concept of memes:

> *Because of the human practice of passing on traditions, customs, knowledge and beliefs ... there is a whole new kind of evolution going on in human beings—a competition not between genetically different individuals or groups, but between culturally different individuals or groups. One person may thrive ... not because he has better genes, but because he knows ... something of practical value.*

Selection between groups happens, and the cooperators grow at the expense of the non-cooperators. Hence the former expand. The history of humanity is one of ever-increasing and ever more complex inter-relationships. Society is not an artificial construct or a tyranny, but the highest form of evolution. What drives progress is increased specialization, and increased trade.

I believe that technology is a semi-independent variable too, part of the great trinity that drives progress; this refinement slots quite neatly into Ridley's argument, since technology itself is a form of culture. It is impossible to explain the explosion of productivity since 1750 purely on the basis of increased specialization and trade; for capitalism to conquer society required new miracles of technology.[4]

Trust

In the end, says Ridley, echoing political philosopher Francis Fukuyama, the crucial human ingredient is mutual trust:

Our minds have been built by selfish genes, but they have been built to be social, trustworthy and co-operative.

He implies that this message goes against the grain of economic doctrine, which may actually be setting back the course of human progress:

If we declare that Smith, Malthus, Ricardo, Friedrich Hayek and Milton Friedman are right, and that man is basically motivated by self-interest, do we not by that very declaration encourage people to be more selfish?

A more functional message, according to Ridley, is that 'the human mind contains numerous instincts for building social co-operation and seeking a reputation for niceness.' And he closes with a passionate plea for what might be termed 'social libertarianism':

The roots of social order are in our heads … We must build our institutions … [to] draw out those instincts. Pre-eminently this means the encouragement of exchange between equals. Just as trade between countries is the best recipe for friendship between them, so exchange between enfranchised and empowered individuals is the best recipe for co-operation. We must encourage social and material exchange between equals for that is the raw material of trust, and trust is the foundation of virtue.

Business collaboration as a means to defeat the evolutionary arms race

Ridley is placing trade, and therefore necessarily business, at the center of the process whereby humans evolve into virtuous collaborators. This is evolution via memes—by cultural transmission—rather than by natural selection. Science is the process whereby we learn about our environment and transmit that knowledge. Business is the process whereby economic information is replicated and used to improve the material conditions of life. The process of constructive human evolution, therefore, requires progressive acceleration and intensification of scientific and business transmission. Individuals and society become progressively more complex and differentiated, increasing the extent of interdependence and 'trade' between individuals and groups of individuals.

Yet, as we saw in Chapter 2, the process really starts a stage earlier, with the formation and replication of business genes—units of economic information. These drive the process of economic development. Like biological genes, business genes derive their power and ability to replicate themselves from combining with many of their fellows and finding vehicles that will protect and incorporate them, including animate vehicles such as individuals, teams, and corporations. Cooperation is at the heart of this process.

Cooperation and competition: twin cherries on a single stalk

Indeed, cooperation and competition are essential complements to business genes, individuals, and corporations alike. Without cooperation between business genes, and between individuals, and between individuals and business genes, there could be no corporations; and without individuals and corporations, there could be no competition. Cooperation between individuals and groups is *how* we compete with other individuals and groups. Over time, cooperation is becoming relatively more important. A sole trader—a single-person business, such as an independent consultant or a prostitute—still needs to cooperate, with intermediaries or bar owners for example. When the typical business unit is larger than one—when the corporation rather than the individual is the modal production unit, an event that is comparatively recent in human history—cooperation becomes not just an incidental requirement, but essential for success.

Cooperation is not just an internal process. Increasingly, it occurs externally, beyond the boundaries of the organization, and beyond geographic boundaries too.

Corporations and cooperation

The modern corporation is unique in economic forms in that it comprises internal (as well as external) cooperation. The essence of a corporation is the free consent of individuals to cooperate to achieve mutual economic ends. A corporation is not a press-gang or an army. The corporation does not own the employees. A corporation is an ever-shifting network of cooperators using a set of economic resources (money, machines, buildings) and a set of common knowledge and technology to provide customers with what they want in exchange for money, which is then shared out between the internal and external cooperators (the employees and suppliers, including suppliers of capital). What makes the corporation work is a web of free cooperation, operating at many levels and with a subtle and unstable texture.

The relative success of corporations is not just a function of how well they compete with (or avoid) each other, but of how well they craft or contrive (organize is too organized a word) the multilevel process of cooperation.

Because cooperation is an instinctive and emotional rather than rational response, there is another currency that is important in eliciting cooperation, which supplements the currencies of cash and reason. The currency of cooperation is emotion, commitment, trust and love.

There are probably two species of successful corporation: those that are excellent at the currencies of reason and cash; and those that are excellent at the currencies of emotion, commitment, trust and love. These correspond to the excellent competitors and the excellent cooperators. It is easy to recognize which companies are in which camp.

The excellent competitors include (in their heyday) ITT under Harold Geneen, GE before Jack Welch, IBM (before the fall), Hanson (before its demerger), Ford, most oil companies, GlaxoWellcome, Monsanto, and any companies that are cold, efficient and low cost.

The excellent cooperators include companies that inspire significant affection in their customers, employees and suppliers: The Body Shop, CNN, Federal Express, Hewlett-Packard, Johnson & Johnson, Levi Strauss, Matsushita, Viking Direct, and Virgin. Many family-owned or family-run companies fall into this category. The personality of the founder(s) often infuses the spirit of firms that are excellent cooperators.

Some companies, admittedly—like McDonald's or Microsoft—seem to have a foot in both camps: they are ruthless and low-cost machines, their employees may generally have ambivalent attitudes to the firms (the Windows 98 project was dubbed the 'death march' by Microsoft employees), but they are popular with customers.

It would be naïve to expect any company to be successful for ever, or to have a very simple explanation for success that is related purely or largely to its processes and culture, ignoring the (somewhat) independent strength of its technology, knowhow, brands, market position and other structural factors. Nevertheless, a reasonable hypothesis is that unusual skill in crafting corporate cooperation is one important ingredient in or route to success. It can also be plausibly argued that skill in cooperation, if not yet obligatory, is becoming increasingly important. As employees become better educated and wealthier, the balance of power between the corporation and the individual shifts in favor of the latter. Gaining the cooperation of the best employees may be the ultimate frontier in competition, at least in knowledge-intensive companies.

Some statistical support for the importance of cooperation is provided by a study undertaken by the Business Round Table and quoted by

Robert Waterman.[5] A 30-year study of 'socially responsible' companies—presumably excellent cooperators—showed that they outperformed the Dow Jones index by 7.6 times!

The theory of co-opetition

In 1996, Barry Nalebuff of Yale School of Management and Adam Brandenburger of Harvard Business School put forward the *Theory of co-opetition*,[6] which aims to combine competition and cooperation. They provide a very useful way for thinking about the two activities:

◆ Cooperation is how we create value: how we create the pie.
◆ Competition is how we capture value: how we grab our slice of the pie.

Drawing on game theory, Nalebuff and Brandenburger say that business is a game where, in order to create value, the company needs to relate to other players. To the conventional categories of customers, suppliers and competitors, they add the 'complementor,' the previously overlooked counterpart to the competitor.

Microsoft and Intel are complementors. Microsoft's sophisticated software packages require ever more powerful chips from Intel. The chips in turn make the software feasible and economic. Sometimes direct competitors are complementors, if, for example, they draw in more customer traffic to a shopping mall. As we'll see later, all networks, like telecommunication systems, transport systems, or the internet, benefit greatly from increased traffic. It follows that the actions of direct competitors, if they enlarge the size of the market, actually benefit all the other competitors.

Nalebuff and Brandenburger provide a mnemonic, PARTS, to help apply game theory to business. PARTS implies Players, Added value, Rules, Tactics, and Scope.

Players. Identify the players and categorize them into customers, suppliers, competitors, and complementors. A competitor is a complementor if your customers value your product more when they also have that other player's product. If they value your product less when they have the other player's product, then it is a competitor. The same concept applies to your suppliers, if they are very important to you. If the supplier is more likely to supply you if the other player is around, it's a complementor. But if you are both fighting over the supplier's scarce supplies, you're competitors.

Added value. Nalebuff and Brandenburger have a nifty way of measuring the value you add to the game. Add up the total value supplied by all the players. Now repeat that, but for all the players except yourself. The difference is what you uniquely add—it's often quite small.

Your strategy, and in particular whether you encourage or rebuff cooperators, can determine how much value there is in the system. Nalebuff and Brandenburger do not say that it is always better to cooperate. They offer a telling contrast between Nintendo's competitive strategy in video games and IBM's competitive strategy in the PC market.

Nintendo's strategy was to ensure that it captured the lion's share of the value in the video games network. It limited the number of games its developers produced to five new games a year, so that quality ruled rather than quantity. It carefully controlled supply to the major retailers, so that there was always pent-up demand and the retailers wanted the games more than Nintendo needed the retail space. It restricted the size of its market somewhat, baking a slightly smaller pie than possible, but it made sure that the pie was highly profitable and that it took most of it. Five years after entering the US market, Nintendo's market value was higher than that of Nissan and Sony combined.

By contrast, IBM invited Intel and Microsoft to help develop its PC. Open architecture and cooperation led to speed of development and a large market. But when other companies copied the IBM PC, it was Intel and Microsoft that cashed in. IBM should have made Intel and Microsoft pay to play, or insisted on cross-shareholdings. IBM brought most to the party, and took away least.

Rules. Rules are an important part of the game and can often be subtly shifted in your favor. But rules can always be rewritten by a creative player who has real value to add. Take the case of advertising agency

Cordiant, Maurice Saatchi, and British Airways (BA). BA was a big client of Cordiant. Maurice Saatchi had been a founder of the company but was ousted. He took with him the key account executives from the BA account. Cordiant was confident that it could keep BA's business, because the account executives had non-compete clauses in their contracts: they couldn't compete for business against their former employer. So what did Maurice Saatchi do? He went to BA equipped with cutout pictures of the executives. What a pity, he said, that these guys can't serve you, explaining their non-compete clauses. So BA went to Cordiant and asked it to lift the non-compete clauses. Cordiant complied, reckoning that it would lose the account anyway and not wishing to forfeit possible future goodwill. End of non-compete clauses in one fell swoop. Yet for decades previously, such clauses had operated effectively in many different professional service businesses.

Tactics. Business, claim Nalebuff and Brandenburger, is often conducted in a fog, where reality is difficult to see and perceptions are all. Game theory can tell you whether and how to lift the fog. When a new product really is superior, they suggest you should take actions that proclaim this. When Gillette launched the Sensor razor, it was so convinced that it had a superior product that it spent $100 million on advertising it to 'lift the fog.' Consumers, faced with such confidence, felt that there must be something worth trying, and Gillette's global sales rose by 70 percent.

But sometimes fog is useful to companies, and enables them to keep a larger share of the pie. A good example is the impenetrably complex fare schedules used by airlines. A misguided attempt by American Airways to clear the fog occurred in 1992, when it introduced Value Pricing and simplified fares down to four categories. Other airlines retaliated, and price realizations crashed. Airlines in the US contrived to lose $5 billion that year.

Scope. Look beyond the boundaries of the game. No business game is an island. Players in one game also play in others. Anticipate and prevent, or at least delay, such invasions. In 1980 a niche toiletries concern, Minnetonka, launched Softsoap, an upmarket liquid based soap. It was a great product, but unpatentable. How could the large toiletries players be prevented from entering? One way was to tie up the entire production of the only two makers of the product pump dispensers for a full year. By the time the majors entered, Softsoap was identified with its category. The brand was eventually sold to Colgate-Palmolive for $61 million.

The Sum of the PARTS is simply analyzing the relationships between all the players in the system. Who needs whom? Who can benefit from the relationships? Who are the actual or potential complementors? Where could cooperation lead to a bigger pie? Even where competition would deliver more value to the corporation than cooperation, adversarial tactics may subtract more value than they add.

The cathedral versus the bazaar

What architecture of corporations—what structure of corporations in society—is likely to facilitate the greatest trade, specialization, innovation, exchange of information, and wealth creation?

Is it better to have a series of large, specialized corporations, each built like a cathedral, with intricate internal symmetry, a sense of their own distinctness and holiness, and a dominating presence; or to have a larger number of smaller enterprises, that exist next to each other, jostling for position and custom, open to each other's secrets and freely exchanging information, like bookmakers in a betting ring, or traders in a bazaar?

Equally, is it better to have a centralized polity covering a wide continent like Europe, North America, or Asia; or to have a series of small, independent states?

The cathedral versus bazaar question arises from a recent debate on the better method for debugging software. Eric S Raymond contrasts the two styles:

I believed that the most important software needed to be built like cathedrals, carefully crafted by individual wizards or small bands of images working in splendid isolation …

Linus Torvalds' style of development [as practiced in his software engineering firm Linux]—release early and often, delegate everything you can,

be open to the point of promiscuity—came as a surprise. No quiet, rever-
ent cathedral-building here—the Linux community seemed to resemble a
great babbling bazaar of differing agendas and approaches (aptly symbol-
ized by the Linux archive sites, who'd take submissions from anyone) out
of which a coherent and stable system could seemingly emerge only by a
succession of miracles.[7]

Linus Torvalds believes that bugs are best fixed by being identified and
later corrected by as large a number of people as possible; the identifiers
and the fixers typically not being the same people. This gives rise to
Linus's law, which is 'Given enough eyeballs, all bugs are shallow.'
Another way of expressing it is: 'Debugging is parallelizable.' In software
debugging, the bazaar is often better than the cathedral, because an
unseen army of collaborators comes to the rescue of the corporation,
and many different corrections and improvements can proceed in
parallel.

The cathedral and *the bazaar*

Linus's law and the processes around it typify the internet culture, where
nobody is in charge and information is freely available. I'll comment on
its implications for networks and economics in Chapter 11. My point in
raising it now is to illustrate a different point, that both cathedrals and
bazaars are useful, competing methods of human cooperation and
exchange. Competition between different groups—organized into corporations with different styles and structures, some more like the bazaar,
others more like the cathedral; and organized into different societies,
some more united than others—drives progress. Sometimes it is useful to
have large scale and more proprietary exclusiveness; sometimes smaller
scale and greater openness. And, in so far as we can generalize on what is
better, the answer appears to be an intermediate degree of concentration
or fragmentation.

Diamond's principle of intermediate fragmentation

Jared Diamond, professor of physiology at UCLA Medical School, has proposed the principle that intermediate fragmentation is optimal in business and society:

> *You don't want excessive unity and you don't want excessive fragmentation; instead, you want your human society or business to be broken up into a number of groups which compete with each other but which also maintain relatively free communication with each other.*[8]

Diamond argues that Renaissance China had a huge technological lead over Europe: for example, in 1400 China had the world's largest fleet, comprising hundreds of ships and total crews of 20,000 men. But in 1432 a new emperor sided with the anti-Navy faction, and decided to dismantle the shipyards and stop sending out the ships. And, because the emperor was in charge of this huge nation, that was that. Diamond contrasts the centralization of decision making in China with what happened with ocean-going fleets in Europe. Columbus wanted a fleet to sail across the Atlantic ocean. Being Italian, he tried to raise support in Italy. Everyone in Italy thought it a stupid idea. So Columbus tried again in France, with the same result. He traipsed from country to country until finally, at the seventh attempt, the king and queen of Spain conceded him three small ships. Europe's fragmentation made Columbus's voyage possible, leading to European colonial empires.

But Diamond observes that extreme fragmentation, as exists in India, favors innovation little better than extreme centralization. And he points out that many industries have a minimum efficient scale, so that an extremely fragmented and geographically protected industry, like brewing in Germany, cannot be as efficient as a more concentrated and competitive one, like America's beer industry. As Adam Smith observed, too small a market constrains specialization. A highly localized market is likely to be less efficient than a continental or global one.

Hence the virtues of intermediate fragmentation, where competition and collaboration coexist. The ideal model, according to Diamond, is

Silicon Valley, which:

> *consists of lots of companies that are fiercely competitive with each other, but nevertheless there's a lot of collaboration, and despite the competition there is a free flow of ideas and a free flow of information between these companies.*

In other words, a landscape with many cathedrals, many bazaars, and free flows of information, competition, and cooperation.

Higher still and higher, the bounds of cooperation rise

Human and business evolution are stories of ever-increasing cooperation. Starting with the family and the small clan, and a series of subsistence, local economies of hunter-gatherers, we have developed into a complex, interdependent collection of larger societies and economies, characterized by ever-increasing degrees of specialization and trade—trade in goods and services, in ideas, in technologies, in roles, and in payments from one group to another.

Specialization and trade are positive-sum games: they create wealth that didn't exist before. They are grounded in reciprocity. They require the habit of collaboration and cooperation for long-term advantage. They require us to look beyond short-term calculations of who benefits from individual transactions; they require, not selflessness, but a long-term, sophisticated view of self-interest, and the habits of daily give and take. They require competition and cooperation: the pursuit of individual and corporate enrichment, and the willingness to enrich other individuals, groups, and society at large. They require social and individual habits of trust, openness, intelligence, and retribution on those who withhold cooperation. They require a balance between high scale and concentration, which aid efficiency at any point in time, and fragmented decision making, which is the best guarantee of dynamism and long-term progress. They require us to have freedom to change, at the drop of a hat, those with whom we choose to collaborate, not because of whim, or short-termism, but because conditions shift and change our perception of where our long-term interests lie.

Competition and cooperation are not opposites; they are necessary complements. Competition is a civilized way of ensuring that we cooperate at the level of society. Because we compete for our livings, we ensure that all workers add social value. The market mediates a potential conflict between how we would like to spend our time, and how others want us to spend it. All business activity requires cooperation, even for sole traders. The advance of business requires ever greater social and material exchange, which themselves require trust, friendship, cooperation, the stick of competition, and the carrot of gain. Ultimately, both competition and cooperation are social virtues and expressions of tolerance and reciprocity.[9]

Selfish genes require humans to behave generously and selflessly

As when the Prisoner's Dilemma is played repeatedly, long-term rewards are reaped by cooperation rather than by pursuing short-term self-interest. The long-term approach requires commitment, valuing trustworthiness for its own sake. Only by behaving in a consistently trustworthy manner— ignoring the opportunities where short-term advantage is available at no cost—will individuals be recognized as trustworthy, which creates the most valuable opportunities. Thus cooperation is based on reciprocity but goes far beyond it. Only by generosity and selflessness, by practicing fairness, is it possible to reap the rich rewards of cooperation.

Cooperation is motivated by selfish genes, but it requires selfless behavior. Humans are programmed by emotion rather than reason, but emotions are highly functional mental devices for guaranteeing commitment. As in later versions of the prisoner's dilemma, the cooperative join forces to cooperate with each other. As Ridley says:

> The virtuous are virtuous for no other reason than ... to join forces with others who are virtuous, to mutual benefit. And once co-operators segregate

themselves from the rest of society a wholly new force of evolution can come into play: one that pits groups against each other, rather than individuals.

The prisoner's dilemma and your career

For individuals, business life, like the prisoner's dilemma game, is best seen as the gradual unfolding of a series of opportunities to cooperate. No transaction is an independent event. The link is your skill at cooperating, the number and quality of people who will cooperate with you, and your reputation. The development of a career in business is not so much about building technical skills as it is about building valuable contacts, people who want to do business with you.

Summary

Modern solutions to the Prisoner's Dilemma provide a good model of how to behave in business (and life). Provided that business is more than a single transaction, which it nearly always is, the dominant strategy is to cooperate, but to punish failure to cooperate in others by withholding your own cooperation until reciprocal cooperation is restored.

Humans have carried cooperation to its highest stage. We are not more altruistic than other animals, but we do live in large and inter-related groups. We have carried cultural learning and knowledge transmission to its highest levels, based on ever-increasing degrees of reciprocity, of sophisticated selfishness.

Business genes drive both cooperation and competition. Cooperation and competition are necessary complements. They can only increase together. But cooperation is the more basic process, and, for each business gene and individual, the more fundamental. Without cooperation, an individual business gene or an individual person cannot produce anything. Deciding which business genes and people to cooperate with is therefore more basic and important than deciding which business genes and people to compete against.

Competition only becomes of paramount importance when it forces itself on you; in its extreme manifestation, when it puts you out of business. Competition then takes you back to the drawing board: it invites you to cooperate in a more successful economic production. Competition takes care of itself. Cooperation requires a prior act of will by individuals.

Cooperation and competition make it possible to specialize, trade, and develop technology. These in turn require ever greater interdependence and cooperation. The groups of business genes and individuals that win are those who cooperate best, and have the best reciprocal cooperators. Trust is the glue that creates a wealthy and well-functioning society.

It's fine to be selfish. But advanced selfishness requires us to be cooperative, selfless and generous. Our selfish genes have implanted our altruistic and cooperative instincts, because this is the way to the highest level of evolution. Short-term selfishness, of the type assumed in conventional economic theory, is relevant only to a small minority of short-term, truly independent, buy-and-sell transactions.

Beyond these transactions lie the important parts of life: relationships—both economic and non-economic—and knowledge, emotions and love. All these transactions require cooperation and trust. Here, displaying short-term self-interest is self-defeating; it tends to undermine trust. The way to get ahead is to have a large network of cooperators, and cooperators find it hard and uncongenial to deal with transparently self-interested people. They will only do so if there is no choice. For a time, greed may drive out cooperation. In the long run, cooperation will drive out greed.

Companies and teams of individuals are above all instruments of advanced cooperation. The best cooperators may inherit the earth—provided that they are also good at satisfying selfish customers.

Ultimately, the advance of civilization and the evolution of mankind and society require ever greater degrees of scientific and economic knowledge and activity, greater specialization, greater trade, greater interdependence, greater competition, and above all greater cooperation. This is the only way to defeat the war of all against all and the prevalence of death and misery over health and happiness—the only route to evolutionary disarmament.

Action implications

◆ *Cooperate with the best cooperators.* Build relationships with the cooperators who possess the blend of business and cooperative attributes that can take your career and business to the highest peaks. Remember that the objective of a career is to build an ever-increasing network of skilled cooperators.

◆ *Build a reputation as someone who creates wealth for others and who is totally trustworthy.* Keep your word, without calculation of short-term gain.

◆ *Always cooperate in the first instance.* Trust others until they prove themselves unworthy of your trust. Only withdraw cooperation from non-cooperators. Punish the latter, but then rebuild mutual self-interest by clear signals: I will cooperate if you will, but only if you will. Demonstrate the disadvantages to others of their failure to cooperate.

◆ *Be willing to 'take turns' in extracting advantage.* Understand that reciprocity is a long-term concept, not one requiring mutual advantage in each individual transaction.

◆ *Develop the daily habits of cooperation.* Teach yourself that cooperation, like thinking and networking, is cumulative and self-reinforcing: it is impossible to deplete your bank balance of cooperativeness by cooperating. Seize all available opportunities to cooperate with useful cooperators, realizing that cooperation builds skill at cooperation, as well as building your reputation, reciprocal obligations, and added skill in cooperation in those with whom you cooperate. Become a cooperation junkie, a cooperation fanatic, an evangelist of cooperation. Always remember: cooperation is the highest from of self-interest.

Notes

1 Robert Axelrod was the organizer. See Robert Axelrod (1984) *The Evolution of Co-operation*, Basic Books, New York.
2 Matt Ridley, *The Origins of Virtue*.
3 See Jared Diamond (1999) 'How to get rich,' *Edge 56*, June 7, www.edge.org/documents/archive/edge56.html.

4   See Richard Koch (1998) *The Third Revolution*, Capstone, Oxford

5   See Robert Waterman (1994) *The Frontiers of Excellence*, Nicholas Brealey, London, Appendix 2. In North America the book is called *What America Does Right*.

6   The title of their book; see Barry J Nalebuff and Adam M Brandenburger (1996) *Co-opetition*, HarperCollins, New York.

7   Eric   S   Raymond   (1999)   *The   Cathedral   and   the   Bazaar*, http://www.tuxedo.org/~/writings/-cathedral-bazaar/.

8   Jared Diamond, *op. cit.*

9   For an excellent explanation of why competition is an outgrowth of and reinforces tolerance, see Ivan Alexander (1997) *The Civilized Market*, Capstone, Oxford. Alexander comments:

'Competition forces political, economic and democratic debate on the advantages and disadvantages of change and rearrangement. This in turns means negotiation, compromise and bargaining. These in turn mean that power is significantly bounded. Containment makes for a habit of tolerance.

'None of these processes are "natural", since nature is not tolerant and does not bargain. [This is a similar argument to that of Richard Dawkins and Matt Ridley: man has evolved beyond nature, if by 'nature' we mean the rest of the animal and plant world.] But, suitably superintended, competition in human society and especially in business becomes a tool of personal and corporate civility.'

Part One Concluding Note

In Chapter 1, we saw that evolution occurs at many levels, but always through the same process of inheritance, experimentation, variation, selection of variants best adapted to the conditions of life, and ruthless culling of inferior variants. Although it takes a very long time, evolution achieves astonishing results, with the luxuriant flowering of all life forms having been derived from one ultimate source. Evolution proceeds through the creation of new variants and new species, which split off from existing forms of life, and through the extinction of less adapted species.

Driving the whole process are genes with similar chemical structures but varying genetic messages. As we saw in Chapter 2, genes are replicators who collaborate with each other to find vehicles—animals and plants—that can help the genes survive and reproduce. We also saw that our species is unique in having devised memes, cultural transmission in the form of languages, art, architecture, science, customs, and ways of doing things, including business enterprise. Memes may offer a way of controlling our genes and creating a new type of evolution.

We also examined the Theory of Business Genes. The most fundamental unit of value in business, similar to DNA, is a unit of useful economic information. This is a meme, but to distinguish it from non-business memes we called it a business gene. Business genes include basic technologies, ideas, products, skills, and individual executives and entrepreneurs. To achieve their purposes, they combine with many other business genes and find vehicles for their replication: more developed technologies, corporations large and small, markets, channels of distribution, and knowledge vehicles such as books, institutes, and university departments. Business progress flows from experimentation, new combinations of business genes, the creation of new business genes, and a struggle for life between the business genes. Corporations are throw-away vehicles for the replication of ever better business genes.

Chapter 3 looked at the experiments of Soviet scientist G F Gause on small organisms. His 'test-tube wars' demonstrated that with limited resources, organisms of the same species will compete to death, but organisms of slightly different species will cooperate to survive. Gause also showed that if one species can invade another without the invaded species being able to reciprocate, the former will become dominant. He further demonstrated that there was a difference between coexistence, where each species can invade the other, and bi-stability, where neither species can invade the other.

We concluded that business genes and corporations should differentiate themselves to survive. They should also find positions where they cannot be invaded by competitors, and preferably positions where they can become dominant by invading a species that cannot retaliate.

Chapter 3 also considered the idea of ecological niches, unique ways of making a living in a specific place in the economy of nature. This reinforced the idea of specialization and differentiation being essential to prosperity.

In Chapter 4, we took a tour into Evolutionary Psychology, and speculated that there is a profound mismatch between the inclinations of our genes and the imperatives of modern business life. Our genes are still geared up for life in the Stone Age, before the invention of agriculture and commerce. Life on the Savannah Plain was hard and frequently threatened. Stone Age man survived by giving emotion precedence over reason, by making quick judgments on first impressions, by banding together in small clans up to a maximum of 150 people, by being friendly, specializing, and cooperating within the clan, by breast beating and mindless optimism, by conforming and herding, by being willing to fight other clans, by avoiding risk whenever there was not a direct threat to life, and by mad scrambling when there was.

While some of these traits—such as a propensity to demonstrate friendliness and work constructively in teams—are functional for modern business, most of them are not. Evolutionary psychology helps to explain the pathology of many large organizations, where there is often a tension between the formal organization and the informal, the proliferation of informal pecking orders, an unhealthy taste for hierarchy and unwillingness to take responsibility, the rejection of negative feedback, a socialistic attitude to rewards, continual bickering between different departments, functions and locations, hostility to out-groups within and beyond the corporation, herding and conformism within it, the suppression of conflict and heresy, the tendency to expect unrealistically optimistic outcomes, avoidance of risk, and panic when things go wrong.

We posited two remedies for the mismatch between neolithic genes and modern corporate life. One is to adapt our behavior: to correct for our natural biases, to cool our emotions, to accept criticism, conflict and contrary views, to use reason and experience to make realistic or even pessimistic projections, to collaborate with out-groups as genuinely as with the in-group, to take more risk than we want, to avoid seeking status or sucking up to it, and to behave reflectively and constructively when under pressure. The other way is to adapt corporate life to our genes, and eschew organizations of more than 150 people.

Finally, Chapter 5 examined the evolution of human cooperation. Humans have learned to cooperate and live in large, inter-related groups, societies of ever greater differentiation and complexity, linked by business relationships and trade. At the root of this process are business genes—customs, ideas, information, technology and skills that can be passed on without sex or natural selection from one human to another.

We saw that competition and cooperation were two sides of the same coin, comprising the links between business genes and their vehicles and the way that wealth can be created by mediating the selfishness of each business gene and individual and using that selfishness to create collective advantage, wealth, and progress. We also saw that cooperation is the more basic process, pre-dating competition and requiring more conscious volition. Progress requires ever greater degrees of cooperation among cooperators willing to take a long-term view of the value of collaboration: cooperators who look beyond individual transactions to an ongoing stream of transactions facilitated by trust and relationships.

In Part Two, we leave biology and turn to physics, in an attempt to understand another dimension of the power laws around us: the properties and interactions of matter and energy.

Part Two

The Physical Laws

Newtonian and Twentieth-Century Physics

Introduction to Part Two

In Part Two, we try to understand matter and energy, and the power laws that drive them. Because physics is about the nature of matter and energy, it clearly has implications for the nature of the universe itself. The power laws of physics rapidly become the way in which we perceive the universe. The way that physical laws operate supplies the template, metaphors and patterns for much else besides—for our models of thinking; our views of society, industry and markets; our view of how individual humans behave and what we are; and even our most fundamental views about God and the coherence or meaning of life itself.

Chapter 6 recounts the wonderful impact on the world of Newton's *Laws of Motion and Gravity*. Meanwhile, in a parallel universe, as we'll see in Chapter 7, Einstein's *Special and General Theories of Relativity* began the process of subverting Newton's physics. Initially just a bizarre curiosity, and yet to find any practical application, the theory of relativity has gradually changed our view of reality, time, and space. It does have interesting implications for business, as long as we don't spend too much time pondering them!

The real hammer blow to humanity's ordered universe came with *Quantum Mechanics*. Chapter 8 tells the story. Although initially the implications of quantum theory brought perplexity and despair, some popular science writers have recently developed a populist 'quantum philosophy' that domesticates the theory and suggests some fashionable implications for individuals, business, and society. We examine these theories with a sympathetic yet ultimately skeptical commentary. We can, however, learn from *Quantum Mechanics* at least one invaluable lesson: that the opposite of a great business truth is another great business truth, and that a 'both/and' mentality can therefore transcend what often appear to be inevitable conflicts.

6

On Newton's Laws of Motion and Gravity

Nature and Nature's laws lay hid in night
God said, 'Let Newton be', and all was light.

Alexander Pope

Newton's impact on the world

Sir Isaac Newton (1643–1727)[1] showed that there are some basic, universal laws, identifiable by precise mathematical relationships, that govern all physical movements on earth or in the heavens. You can predict what will happen if you put a ship on the sea or roll a penny down a chute. With Newton's three **Laws of Motion** and his **Law of Universal Gravitation**, you can build bridges, fly planes or send men to Mars, and be pretty confident of the results.

Newton's *Philosophiae Naturalis Principia Mathematica* was published (in Latin) in 1687. Here is the English translation of his three laws of motion:

Law I   Every body continues in a state of rest, or of uniform motion in a
right line, unless it is compelled to change that state by a force

impressed upon it.

Law II   *The change in motion is proportional to the motive force impressed:
            and is made in the direction of the right line in which that force is
            impressed.*

Law III   *To every action there is always opposed an equal reaction; or the
            mutual actions of two bodies are always equal, and directed to con-
            trary parts.*

Newton's first law is that things keep moving forward in a straight
('right') line unless interfered with. This is a restatement of Galileo's law
of inertia, that bodies remain at rest or in constant motion, except when
moved by an outside disturbance.

The second law is that force (*F*) is directly proportional to the change
in momentum that it generates. Twice as much force will cause twice as
much change in an object's momentum. Newton provided an original
definition of momentum—mass times velocity—where mass (*m*) is the
'quantity of matter' in an object. Change in velocity is the same as acceleration (*a*). Hence Newton derives his famous formula: $F = ma$ (force
equals mass times acceleration).

The third law of motion, Newton's most original insight, is that every
action produces an equal and opposite reaction. If similar objects collide,
they bounce off with equal force. If an object's motion is disturbed (its
momentum changes), then the motion of another object must also be disturbed so that the 'aggregate' momentum is unchanged. The second disturbance must be precisely equal to the first, but in an opposite direction.

From these three laws, and Galileo's law of uniform acceleration,
Newton arrived at the concept of gravity. An object falls to the ground
and its momentum increases. Newton's three laws say that some force
must be responsible for acceleration, and this force must be constant if (as
Galileo showed) acceleration was constant. This force must therefore be
gravity. He concluded that the force of gravity on an object is constant
and directly proportional to the mass of the object. Hence Newton's law
of gravity (also known as his 'inverse square rule') that *between any two
bodies, the gravitational force is proportional to the product of their masses, and
inversely proportional to the square of the distance between them.*

The most important external force, whether on earth or in the sky, is gravity. Newton's laws of gravity—together with the data and laws provided by astronomer Johannes Kepler—finally established that the earth revolves around the sun and not the other way round. Newton showed that the planets' movements around the sun fitted the equations of gravity, but required (for the calculations to work) slightly elliptical orbits around the sun. Planets try to 'go straight,' but gravity forces them into a curve. The orbits can be calculated if we know the mass of the planets and the distance (and hence the inverse square of the distance) between them.

Amazingly, what holds the planets in place is the same as what makes apples fall off trees: gravity. Once gravity is appreciated, the heavens can be seen to move, not randomly or in a complex pattern, but just like clockwork. Gravity bends the movement of planets and moons, and the extent of gravity is a function of how close an object is to the sun or other force of gravity, and the relative mass of the object and the sun.

In the Overture, we saw that Newton was a synthesizer of centuries of scientific insight, and in many ways he became a symbol of a new scientific perspective: in praising him extravagantly, contemporaries and later writers were really celebrating a new sense of intellectual coherence and liberty, in which Newton played a leading but well-supported role. Nevertheless, he and his colleagues changed the world. Newton's amazingly economical theses on motion and gravity showed how a few simple rules, worked out from first principles and validated by mathematical proofs, could have universal application. The popular implication was that the world was predictable and controllable by scientists and engineers. This was deeply reassuring and inspiring, and remains so today, even when we know that Newtonian physics is incomplete and very slightly inaccurate.

Are Newton's power laws 'old hat'?

It is common these days to denigrate this heritage and point out that the 'Newtonian,' mechanical and rational view of the world has huge and distorting gaps. We'll come to these in the next few chapters. But the 'reaction' has slithered too far towards fashionable but glib notions, taking for granted or ignoring the terrific value provided by Newtonian tools and concepts. Where would we be, for example, without the simple ideas of profitability and the power of comparing a few simple and universal numbers—return on sales, return on capital, and the internal rate of return on a project or investment? Would we be wealthier with or without the idea of budgets and the practice of reviewing them? Is it sensible, or just old-fashioned, to look for the few causes determining success or failure, to identify common characteristics, and to test our theories with numbers? And would we really be better off without the machine metaphor and the mechanical view of life, the universe, and everything?

A thought experiment: a world without business numbers

Imagine that you used time or space travel to visit a sophisticated economy that had thrown away or never used the Newtonian tools of business. I am going to be generous, and give this imaginary world the benefits of technology, teams, corporations, computers, stock exchanges, and nearly all of the apparatus of modern economic life (even though I could plausibly claim that these mainly derive from Newtonian methods). But this imaginary world does not have accounting systems or the ideas of return on capital and discounted cash flow. For all their defects, imagine being able to make investments using these methods, selecting by apparent magic the most profitable and promising companies. How long do you think it would take you to become the richest person on the planet?

Perhaps the reason that business writers find it more attractive to rubbish the rational, machine-based, analytical approach to business than to celebrate it is that there is little to add to the rational school. A long line of management thinkers, from Frederick Taylor to the Harvard Business School writers, culminating in the 'microeconomic' analysis of business positions by Professor Michael Porter, has mined this seam so well that there is apparently little more to say. Taylor's *The Principles of Scientific Management* came out in 1913, and Porter's ground-breaking *Competitive Strategy* in 1980. Between these two dates, virtually everything of a Newtonian and rationalist nature, that could be said about business relationships and what determines profitability, was said and said well.

Yet not quite everything. The golden age of Newtonian business analysis may be over. However, in my view one key insight has been insufficiently appreciated and another overlooked altogether. The rest of this chapter explores these insights.

Action and reaction

First, a short note on the one that has been noted, but perhaps not given due weight. Newton said that action and reaction are equal and opposite. Loosely interpreted, this means that any powerful development, whether a school of thought, a new technology, or a new market, will bring in its train an equally significant opposite development. We can see this clearly in the realm of ideas or proposals for organizing society: capitalism produces socialism, the experience of trying to make socialism work produces a neo-capitalist revival, and the primacy of global free markets will doubtless produce another powerful backlash. I've just commented on the same theme in business ideas: because the rationalist, Newtonian school of business thinking has been so powerful, it has conjured into existence an opposite, systems-based school. Ideas and actions break inertia, and once inertia is broken, reaction is inevitable.

Applied to business and markets, this simple insight is an almost infallible guide to what is around the corner. Never mind the important new market, think about the one after that: its opposite. Thus Henry Ford standardized and mass produced the modern car; the next stage was General Motors' production of a range of different models, pitched at different levels. Once an effective, publicly owned healthcare system is introduced, this creates a large demand for private healthcare: a counterintuitive but clear relationship. The same applies to education or any other service. Once there is a mass market in anything, there is a demand for the opposite: niche markets tailored to particular customer groups. Once there is broadcasting, there is narrowcasting.

The beauty of the dynamic is that it works both ways. Once there are expensive, luxury products, there is a demand for a stripped-down economy version. Are luxury hotels, planes, and boats the norm? Then mass tourism will emerge. Are package tours prevalent? Then the market for something superior and tailored will reemerge on a large scale. Do businesses find constant traveling an expensive nuisance? Then video-conferencing and the internet can be substituted. Are the latter important? Then the virtues of face-to-face communication will be rediscovered. Are doctors, nurses and hospitals an important part of modern society? Then alternative medicine will thrive.

New technology and markets often fail to have the drastic impact on their predecessors that is confidently predicted. Personal computers breed prolifically, and gurus pontificate on the 'paperless office,' yet paper and photocopying persist. The phone, the fax, the internet, and video-conferencing do not stop meetings or travel or book reading.

Nevertheless, this observation is a slight *hors d'oeuvre*. Now for the main course.

The gravity of competition

There is one direct parallel between Newton's laws of motion and the nature of business that has been overlooked: that between gravity and competition. I think I can establish that this has definite metaphorical and conceptual value; but it may even be possible, in the fairly near future, to use Newton's principles to establish a quantitative (inverse) correlation between exposure to the largest competitor and the return on capital earned by a business in a particular competitive arena. Let me first explain the concept, using the metaphor of gravity.

Competition is a grave affair. It is the economic equivalent of gravity. Just as gravity depresses objects, and stops stars going straight, so competition depresses returns on capital. Margin gravity depresses managers and investors. The extent of margin gravity is proportional to the proximity and power of competitors. Weak gravity indicates distant or tangential competitors. Strong gravity implies close, in-your-face competitors.

Black holes

A black hole is formed when a huge, heavy star burns up all its fuel and collapses so far that nothing can escape from it: no light, nor any other kind of signal, can emerge. Black holes only occur if the star is really heavy: about three to six times heavier than the sun, according to Einstein's theory of general relativity. When such an enormous star dies, a black hole collapses all the space around it. The resultant mega-gravity curves the adjacent space to a fantastic degree; the gravitational pull approaches infinity.

The business 'black hole' is a useful metaphor describing competition so intense and head to head that margin gravity approaches infinity. No profits or positive cash flow can escape from a black hole.

Competition-free zones

There are places that are free, or virtually free, from margin gravity. These are spaces where competition does not operate, where margins are limited not by competition but by what customers can afford, and by the distant hiss of competition from all other products and services clamoring for the customer's wallet. These non-competitive spaces, where gravity doesn't work, are the most beautiful places in the economic universe.

Corporate gravity: how near and large is your competitor?

What stops your firm moving forward in a straight line, and achieving its business plan, is pesky competitors. Competition is the corporate equivalent of gravity; it brings you down to earth. If you bang up against a competitor, you will both be thrown off course.

The force of competition is a function of two things: the relative size (or, more precisely, mass) of your most important competitor, and your distance from that company.

A small and under-resourced but very adjacent competitor may cause you more trouble than a huge, rich and successful corporation that is only marginally interested in your markets.

What is size?

'Size' (or mass) really means the resources that a competitor can devote to a market; it implies profitability, skill, and fitness to serve customers more than simple size in revenues. Size can also mean strength of balance sheet, ability to get new capital, a firm's price/earnings ratio, and the strength of its reputation, brands and relationships. Until we can measure these aspects of 'size' quantitatively, a good proxy may be one that has been used in business since it was invented around 1970 by the Boston Consulting Group, relative market share (RMS), which is your revenues in the business segment divided by the revenues of your largest competitor. (If this number is over 1.0, it implies that you are the segment leader; if less than 1.0, that you are smaller than the leader; if exactly 1.0, that you are co-leader with one of more rivals of the same size.)

What is distance?

'Distance' is the extent to which the competitor is close to your own customers. A distant competitor is one whose focus is elsewhere. A close competitor is one that has the same target market and the same approach.

Distance can usefully be envisaged as a series of ever larger squares, with the corners being:

◆ customer type
◆ product type
◆ geography
◆ type of value added provided (e.g., research and development, production, distribution, marketing, sales), either an integrated operation or a specialist by stage of value added.

If you and a competitor are serving the same type of customer with the same type of product in the same geographic area, and providing exactly the same type of value added, there is virtually no distance between you and the other company. You are on top of each other. The converse also applies: the distance is huge if the customers, products, geographic markets, and type of value added are all different. A crude scoring system to calculate distance is to score any business segment on these four dimensions as follows:

1 = identical or very similar
2 = adjacent, close, similar but with a few differences
3 = neither very close nor very distant, considerable overlap
4 = not very similar or close but some overlap
5 = distant, dissimilar, fundamentally not the same.

You can then calculate distance on a scale of 1 to 625 by multiplying the four numbers (the scores on the four dimensions). For example, competing with the same customer type and product in an adjacent geographic area but with a very different value-added focus would produce a score of $1 \times 1 \times 2 \times 5 = 10$). A score of 8 or below indicates dangerous proximity, and anything below 20 indicates close competition.

After making the calculation, your nearest competitor may not be the one you first thought of. In this case, you'll need to calculate the 'size' of the competitor you now think is closest. Remember also that the calculations shouldn't be made at the overall corporate level, but within each separate business segment, where the competitors or customers or profitability or strategy are different.

Can we measure corporate gravity?

In Newton's world, gravity is strong and unavoidable if you are close to a large object. Gravity is still important if you are close to a smaller object, or in the general vicinity of a large object. The effect of gravity in taking you off a straight line can be precisely measured if you know the mass of the object and its distance.

In the corporate world, the gravity of competition is strong if you are close to a 'larger' competitor. But what is the gravity itself? It is not, demonstrably, the ability to serve customers well. This is unaffected by close, head-to-head competition; indeed, from the customers' viewpoint, close competition is usually beneficial. Instead, the gravity is the margin that can be earned by each corporation. What depresses margins is the gravity of competition.

If we could reliably measure the 'size' and 'distance' of competitors, as described above, we could calculate what margin we would expect in any particular market segment. We could then compare this normative profitability to the actual profitability experienced, and see the extent of the correlation. This would almost certainly be a much better measure than market share or relative market share alone, yet carefully segmented studies of the latter, as in the PIMS (Profit Impact of Market Share) projects, have shown a significant correlation. We might therefore reasonably expect the better measures to show a much tighter correlation.

For the present, however, this is just speculation; we shall have to wait several more years (or centuries?) before the business schools—or, more likely, consulting firms—catch up with Isaac Newton. But the intuitive appeal of the concept should be clear. To enjoy high margins, you must be a long way away from competent competitors. To avoid margin gravity, you must avoid close competitors, whether large or small, *regardless of whether you are larger than them*. If you are larger than them, you will hurt them more than they will hurt you, but you will still be hurt. Margin gravity gets increasingly serious as a function of the 'size' (competence and resources) and 'proximity' (similarity of target market and method of serving it) of competitors.

Margin gravity is not linear— returns rise sharply as you avoid competitors

In Newtonian astronomy, what matters in calculating orbits is the mass of the objects and the inverse square of the distance between them. It is a logarithmic rather than a linear relationship. The same is likely to apply to the gravity of competition. We may be fairly confident that, if we can increase the distance between our firm and its most important competitor, or if we can increase our relative size, the impact on profits will be more than linear. This makes sense intuitively, because margin gravity should decrease more than proportionately to an increase in distance or a decrease in the relative size of the competitor. But is there any quantitative, empirical support for this hypothesis?

It so happens that there is. In the 1980s, the consulting firm of which I was a co-founder measured the relationship between return on capital employed (ROCE) and relative market share (RMS) for its clients in many thousands of business segments in many different countries. We found that there was a strong correlation between high relative market share and high profitability. We also found—and this is the key point for margin gravity— that *the relationship was more than linear.* A 10 percent improvement in relative market share produced a greater than 10 percent increase in profitability. It followed that the greatest benefit in profit terms usually came from increasing market share in those markets where the client was already very strong. This is very similar to the concept of 'increasing returns to scale' promulgated by economist Brian Arthur at about the same time.[2]

Similarly, when you move further away from competitors, returns should increase more than proportionately to the distance moved.

How do you increase size relative to your main competitor?

One simple answer is by increasing relative market share. Your sales have to increase faster than the competitor's. One way to do this is to find new customers or sell more to existing ones, at a rate faster than the rival. The other way is to increase your relative rate of retaining the customers you already have.

But, as we've seen, there is more to 'size' than the simple definition of relative market share. 'Size' also means increasing your skill at serving customers, your reputation, and your financial resources. These are what underpin the sales: commitment to the market, understanding of it, and ability to court popularity among customers through better products or service, faster delivery, superior marketing, or lower prices.

How do you increase distance from your main competitor?

In principle, it's simple. You differentiate yourself from the competitor by increasing the differences in your stages of value added, and in the customer types, product types, and geographic regions that you serve.

The only caveat is that moving away from one competitor may bring you closer to another. If you are much larger than the new nearest competitor (or, more precisely, if your relative market share versus the new competitor is higher than it was against the previous closest competitor), this may not matter so much. If the new main closest rival is larger in the relevant segment than the old one, this may be a step backwards. Therefore you need to iterate the possibilities until you can find a way of both increasing your distance from the nearest competitor and increasing your relative size versus that company.

Escaping corporate gravity

There are clear and important lessons from the concept of corporate gravity:

◆ The best way to avoid being thrown off course is to avoid competitors.
◆ Competitors exert downward pressure on margins—that is, margin gravity—according to their relative 'size' and 'distance.'
◆ Resources should be concentrated in segments that exist or can be created where competitors are as far away and as insignificant as possible.
◆ Segments with little or no competition are qualitatively different from those with close competition.
◆ These 'competition-free zones' are not subject to significant margin gravity. The only thing that distinguishes them from normal business segments is the absence of relevant competition. The customers may be the same as in other business segments. The technology may be the

same. The suppliers may be the same. The executives may be of the same ilk. But in competition-free zones, the laws of economics and cosmology are light years apart from the laws in normal business. The value of these zones is truly astronomical.

◆ In competition-free zones, the constraint on margin is not direct competition. It may not even be the wish to deter potential competition, or avoid the wrath of regulators. The real economic constraint is simply the size of the market and its price sensitivity. Margins should find the level at which the value of the future stream of earnings is optimized. This may be difficult or impossible to calculate, but returns may be extremely high. A persuasive argument may be that the size of the future market may be maximized by having *higher* rather than lower margins, if a chunky portion of the margin is then reinvested in improving the product and service and in marketing it more effectively.

◆ In normal competitive markets, don't expect any action or innovation to be ignored by the competition. Work out in advance your reaction to the competitor's reaction—and to the subsequent reaction, and so on.

◆ Improvements that anyone can copy—even that only one competitor can copy—will benefit customers, but not your firm or its investors.

◆ All energy should be devoted to improvements and innovations that increase your distance from significant competitors. This should not be thought of in the conventional terms of having a 'competitive lead.'

◆ A lead is something that connects two parties. You shouldn't be seeking a lead. You should be seeking to escape from the gravitational pull of competitors. An ever-growing distance from competitors is easier to achieve than an ever-growing lead over them, and also much more valuable. To gain a lead, you do things better. To put distance between you and them, you go in a different direction—preferably the *opposite* direction to theirs.

Summary

By seeing the universe as a great machine, and demonstrating how movement on earth and in the heavens followed precise mathematical rules,

Newton showed us how to control our fate and advance industry and science.

A powerful new trend or market will tend to generate its opposite.

There is a close parallel between the method used by Newton to measure the effect of gravity on orbiting bodies in space, and the method we can use to think about the effect of competition on our profits. Gravity is a function of how close an object is to another, and the relative mass of each. Competitors act on a firm in the same way that gravity acts on all objects. Competitors stop your linear progress towards your goals. Competitors curve the space around your actions and depress the margins that customers would otherwise let you enjoy. Competitors cause margin gravity.

Margin gravity is greater the closer you are to a large and highly competent competitor. Margin gravity is greatly reduced if the competitor is distant, small, and incompetent.

A competitor is close if it makes its living in the same way: has the same customers, employs the same technologies and type of people, has the same type of physical assets and the same methods of distribution, has the same suppliers, the same core method of adding value, the same strategy and mindset, and the same priorities as you. A competitor is distant to the extent that all these dimensions are different.

A competitor is 'larger' (has more mass) if it has more resources than you: a bigger balance sheet, higher cash flow, greater access to new capital in the form of debt and equity, a higher rating on the stock market, better brands, and/or greater skill, market share, revenues, attractiveness to key personnel, and ambition.

When Newton calculated the effect of gravity on orbiting moons or planets, he used the inverse square of the distance between them. The gravity reduces with the square of the distance between the objects. Gravity drops off more than proportionately when distance increases. A similar effect is likely with competition. When the distance or difference between competitors increases, the extent to which margin gravity operates falls off dramatically.

In some segments, the gravity of competition is so great that no profits or positive cash flow can emerge. We may call these 'black holes.'

The opposite type of segment, the ideal place to be, may be called a 'competition-free zone.' These zones are likely to be very profitable—and can probably be made even more profitable without damaging your competitive position.

Action implications

◆ *Escape from the gravity of competition.* Systematically increase the distance between yourself and competitors, on the four dimensions of type of value added, product type, customer type, and geographic markets served. Work out how you can do things differently from your closest large competitor, in order to increase the distance between you.

◆ *Where you are already the leader, increase your relative size, and the degree of difference, between yourself and all significant rivals.*

◆ *Focus all your energies, cash and people on business segments where you are already large and a long way distant from any competitor,* or where you can reach this state.

Notes

1   Although discussing it is beyond the scope of this book, Newton is a fascinating character, partly because of the contrast between the conventional view of him as the world's most influential scientist ever, as the father of modern empirical science, on the one hand; and on the other, the more complex reality, that he was a brilliant synthesizer, but very far from a rationalist, a man who spent most of his later years poring over the Bible, inventing bizarre theological fantasies, and tending bubbling cauldrons to discover the secrets of alchemy. For good descriptions of the conventional view of Newton, see John Simmons (1996) *The 100 Most Influential Scientists* [where Newton tops the chart], Carol Publishing Group, New York; and the more populist and entertaining Melvyn Bragg (1998) *On Giant's Shoulders*, Hodder and Stoughton, London. For a scholarly and highly readable account of the complexities of Newton's character and intellectual influences, see Michael White (1998) *Isaac Newton: The Last Sorcerer*, Fourth Estate, London.

2   See Chapter 11 for a discussion of increasing returns to scale and their role in the so-called new economy.

7

On Relativity

*No one can recall without a thrill his first encounter with Einstein's
Carollian world where space–time is curved, a fourth dimension, and honest witnesses blithely disagree on the most elementary questions of what
happened when and where.*

Stephen Hawking

Farewell, clockwork universe

The great clockwork universe of Isaac Newton—the rational, mechanical world where causes and effects can be calculated and where solid reality underlies our steps—is the world we think we inhabit, and to a large
extent we are right. Yet the twentieth century introduced a new form of
physics demonstrating that, at least in the world of very small and 'fundamental' matter, the world is a great deal more wonky, unpredictable and
complex than Newton ever knew.

Newton and his successors living before the twentieth century
believed in absolute space and time, and in our ability to measure and
control all aspects of the machine called the universe. It was a beautiful
dream: the idea of science leading us to be able to control everything;
including, Freud added, ourselves. Yet the dream was shattered, first by

Albert Einstein's insights into relativity, and secondly, and even more dis-
turbingly, by Niels Bohr and the other great discoverers of quantum
physics. In the world revealed by relativity and quantum physics, nothing
is fundamentally real, measurable or controllable. Nothing is what it
seems, and really bizarre things happen at the heart of science.

That is why we should have a rudimentary grasp of relativity (the subject of this chapter) and quantum theory (the next chapter). We businesspeople still live in the Newtonian world. This is not bad. If you didn't
believe that your actions could lead to positive results, if you stopped
measuring cause and effect, threw away your budgets and financial statements, or stopped quantifying the effects on profits of competition and
customer retention, you'd be very lucky to be successful.

I'm not going to ask you to throw away the habits of a lifetime, your
Newtonian, mechanical business models. They have great value. In the
last chapter I asked you to hone those models, to understand the impact
of competition on profits in a much more rigorous, Newtonian way. But
we're about to see that this is only part of the picture. There are things
that we cannot control or even understand through Newtonian thinking. For example, we think of 'organizations' in a classic Newtonian way,
and we deceive ourselves into thinking that we can easily control and
'organize' them. Relativity and quantum physics add a whole set of new,
fresh and liberating insights into life and business.

Reflect now, as you read, on the differences between the world you
thought you knew and the weird world that you *really* inhabit.

Einstein's special and general theories of relativity

Albert Einstein (1879–1955) was the first scientist to establish that there
were fundamental truths about the physical universe that had eluded
Isaac Newton. Einstein's theories of relativity provided a new basis for
understanding space, mass and energy. The special theory came first, in
1905, showing how atomic and subatomic particles work. The general
theory, published in 1916, rested on Einstein's insight that acceleration is

precisely the same as gravity, and made modern cosmology possible. Without Einstein's theories we might still not have transistors, electron microscopes, photoelectric cells or computers; nor nuclear bombs and nuclear power.

The warping of time and space

Relativity, particularly the general theory, is very hard even for physicists to understand. So I am not going to try to describe it. Instead, I shall note some of the most important consequences of Einstein's thinking:[1]

◆ He postulated that light is a stream of particles, whose energy could be calculated ('Planck's constant' later proved the point). These particles of light came to be called 'photons.'

◆ The special theory of relativity contradicts our intuitive views of time and space. Einstein says that nothing (such as a signal) can travel faster than the speed of light, and that light's speed, which had been computed, does not change as a function of the velocity of the observer. Yet it follows that no two observers, going at different speeds, will agree precisely when an event occurs. Time and space, therefore, are not fixed, absolute quantities. It also follows from the special theory that *where* an observer is determines *when* he or she thinks something has happened. You can never precisely say, 'This happened at this time and this place.'

◆ The special theory can be applied, as Newtonian physical laws could not, to predict what happens at the subatomic level. As Einstein said, 'the mass of a body is a measure of its energy-content.' Hence his brilliant equation, $E = mc^2$, where mass (m) is expressed as an amount of energy (E) when multiplied by the square of the speed of light (c). The special theory was very helpful in elaborating quantum theory, which arrived courtesy of Max Planck and Niels Bohr in the early years of the twentieth century.

◆ The general theory deals with gravity and corrects Newtonian physics. It extends the special theory to take in systems that are accelerating, like bodies in space. As a result of the general theory, therefore, we enjoy all the insights from twentieth-century cosmology, including the expanding universe and black holes.

◆ Einstein had the vision of someone falling inside a plummeting lift that had broken its cables: the person would free fall inside the lift, just as astronauts in orbit around the earth feel weightless because they are 'falling' towards it. Einstein posited that gravitational force and the force of something that is accelerating are indistinguishable. There is no way to tell the difference between gravity and acceleration, so there is no real difference between them. Gravity is not, as was previously imagined, the force by which all objects are attracted to each other. Rather, gravitation is the warping of space and time by physical mass. Space is curved, and the elliptical orbits of the planets can be calculated precisely using the general theory of relativity.

◆ It follows from the general theory that time is not independent of space. Time looks and acts like a fourth spatial dimension, and can be warped by gravity. Given that the speed of light is a constant, time and space become a united frame of reference. Einstein refers to events in a four-dimensional 'space–time continuum.'

◆ Even more spookily, Einstein queried whether 'space' and 'time' were realities of nature rather than being simple psychological effects. As the shape of 'space–time' depends on gravitation—requiring material bodies—space and time would be meaningless without bodies. Einstein stated:

It was formerly believed that if all material things disappeared out of the universe, time and space would be left. According to the relativity theory, however, time and space disappear together with the things.

Einstein makes time part of the physical universe

Einstein's greatest legacy is probably his elevation of space and time to dynamic things on which we can experiment. Time, he says, is itself part of the physical universe: it is relative, not absolute. So instead of the three dimensions of space, we should think of the four dimensions of space–time, time being the fourth. Space and time can be changed depending on how fast you travel and how much gravity you experience. Space and time; energy and mass—these are linked together. When the sun shines it is converting some of its mass into energy and light; it is a nuclear reactor. Einstein's insights led on to nuclear power and nuclear bombs: his reaction to Hiroshima was: 'If I had known they were going to do this, I would have become a shoemaker.'

Can relativity be applied to business?

Einstein's theories are difficult enough to apply in science; can they really be applied to business? In a strict sense, the honest answer is 'no.' But ask instead the question: Have Einstein's theories of relativity usefully influenced our view of life and truth, our modern world view? The answer, surely, is 'yes.' So, as long as we realize that there are no Einsteinian equations to support us, and that we are not strictly applying relativity theories, I think that it is fair to suggest that there are two ideas, broadly derived from relativity, that are extremely instructive, namely:

◆ Time is not a separate dimension in business; rather, it is integral to competitive advantage.
◆ We need a 'relative,' not absolute, view of our world, including our business world.

Time: at the heart of competitive advantage

Einstein incorporated time into the physical universe. Time, he said, was not 'other,' an objective external dimension against which everything else should be measured. Instead, time and space were linked together as dynamic things on which we can experiment. Physicists and astronomers today often talk about 'space–time' as one concept.

Curiously enough, this insight about time was implicit in folklore many years before Einstein. In the nineteenth century people said that 'time is money.' In other words, time is intimately bound up in the process of production and can substitute for, or be substituted for, monetary value. Today we might add that 'time is product' and 'time is service.' The same product or service offered faster (or slower) is not the same product or service; it is different and better (or worse). A new product or generation of improved product that is offered to customers after a one-year gap rather than the previous two years is an acceleration of value delivery to customers.

A major challenge for any business is to integrate time into the product or service offering. The objective is to deliver the product/service faster than you used to, and faster than your rivals. There are various techniques for doing this—we'll come to a few in a moment—but the biggest obstacle and opportunity is mental. We think about time as external, as another dimension, even as the enemy. We do not, but should, think about 'product-time' or 'service-time' as internal, part of something that we offer, a crucial dimension that is intrinsic to our way of doing business.

The first mental challenge, therefore, is: *think of time as a friend, a resource, a colleague, and as part of the value you offer customers.* This really is a challenge. We're used to thinking of time as a constraint or an unwelcome intruder; here we are doing something and at last getting somewhere and … oh dear! Time is up! In the words of Andrew Marvell's great poem:

> But at my back I always hear
> Time's wingèd chariot hurrying near.

This is the wrong attitude, yet we should realize how deeply it is woven into the fabric of our thought.

Time heresy: time is abundant, time is available

How can we change our mental map of time? Here are a couple of related thoughts that are profoundly and demonstrably true, and that invalidate our usual view of time. First, *there is no shortage of time*; rather, we are positively awash with it. I may need some time—which we enjoy in abundance!—to persuade you of this, so I will pass swiftly on to point two, which I think will be accepted more readily: *very little of what a firm (and by inference, an executive) does adds a great deal of value to customers.* Most of the value that is added comes in short bursts, oases of productivity, surrounded by a desert of low value processes. Einstein is an extreme example. How much time did he need to devise $E = mc^2$? What was the value/time relationship? Not infinite, but a very big number!

This second point has actually been proved quantitatively by a very large number of surveys, especially those associated in the late 1980s and the 1990s with the techniques of time-based competition and reengineering (also known as business process reengineering or BPR). I don't want to go into techniques yet, or in much detail at all, because I want you to focus on the principle, on your mental map, on how you think about business. But Mark Blaxill and Tom Hout of the Boston Consulting Group sum up their evidence from a huge amount of client work as follows:

> *Typically, less than 10 percent of the total time devoted to any work in an organization is truly value-added. The rest is wasted because of unnecessary steps or unbalanced operations.*[2]

What happens is that the high-value work takes a small amount of time, and there are huge gaps before high-value service to customers is resumed, because firms organize themselves to follow their own procedures instead of having the work flow dictated by the customer's needs. Most of the time a product or service is being produced is time spent waiting, usually for some other executive or part of the organization (and occasionally customers themselves) to respond or do something. Delays come from procedural constraints—always able to be eliminated—quality problems—ditto—and structural difficulties, which can be dealt with by redirecting the flow of work; that is, by changing the structure. Very rarely are firms' structures designed with the objective of speeding the product or service to customer. When they are, there are nearly always large cost and quality improvements as by-products.

So let's come back to the first point: time is plentiful. I know this sounds nonsense, or at least paradoxical. We're all stressed, we're all busy, we're all interrupted, we're all trying to 'manage' our time better, to dole out such a scarce resource as parsimoniously as possible. Yet herein lies the problem, and the answer.

Hold on, relax, and think for a moment. *If only 10 percent of our time is really used to great effect, it follows that 90 percent isn't, and that this time is available for high-value activity.* If we take the numbers literally, we could double our high-value activity and still have 70 percent of our time left to waste. This goes both for individuals and for the corporations in which we work. So we shouldn't be worried about shortage of time; this is an illusion. We don't need to speed up. What we need to do is stop spending our time in low-quality/low-output ways.[3]

Integrating time into your product and service

Einstein's challenge is the difficult, mental one: think of time, or the reduction of time, as part of what you offer customers. Think product-time. Think service-time. It's all part of the same thing. Never think 'product' or 'service' independent of 'time.' Time is a key dimension that must be embraced to achieve success.

See if you can maintain this way of thinking for an hour, for a day. It's tricky! It really is a revolution in the way we think. Believe me, this is 90 percent of the battle. If you're wasting your own time, and that of the customer, there is a wide buffer zone of improvement available, not from speeding up what you do now, but from only doing things that are important to the customer and from organizing around the customer.

Just a few hints about what to actually do:

◆ Measure the time it takes to do things for customers. The time from taking an order to fulfilling it is the most important 'thing.' But there are others: for example, the time to introduce a new product or service; the time to provide after-sales service, and respond to questions or complaints; the time to incorporate an important customer suggestion into a product or service, and so on.

◆ Find out the dimensions of time and time saving that are most important to customers. Then find a way of delivering on these aspects two to three times faster than you have done historically, and two to three times faster than your fastest rival.

◆ Identify separately the procedural, quality and structural issues that are wasting time.

◆ Map out the process of delivering the product or service to the customer, and where the time is being used. Identify gaps that interrupt the flow and eliminate them.

◆ When you have improved your delivery time to customers by two to three times, focus your marketing and selling effort on selling more to your existing customers for whom the benefit of faster delivery and greater responsiveness is greatest; and on finding new customers who will also value the time benefit significantly more than other customers.

◆ Measure customer retention; that is, the proportion of customers who repeat purchase from you. Ensure that you raise the customer retention percentage each year, especially the retention of your most valuable and profitable customers. Use your time advantage to improve the customer retention.

A relative world view

Way back in 1905, in explaining what became known as the special theory of relativity, Einstein took the first step—later steps coming courtesy of quantum theory, and to a much greater extent than Einstein liked—towards undermining our view that anything is absolute and fundamental. Einstein said that the speed of light was a constant. It follows that if two different observers are traveling at different speeds, they won't agree on the precise time that anything happened. Practically, the differences are very small indeed. But the mold of absolute measurement and absolute reality was broken, once and for all, by Einstein's insight.

It was the same with geometry. The old view was that 'parallel lines meet only at infinity.' This is no longer true. Einstein's theory of relativity holds that space is curved, and that parallel lines *do* meet this side of infinity. A long way off, yes, but the difference in world view is stunning. The cat is out of the bag. Everything is relative.

You can see the difference between the seventeenth- to nineteenth-century view and the twentieth-century view almost everywhere: in science, in literature, in popular songs, in art. The old view was one of certainty, predictability, and absolute confidence. The heritage bequeathed by the twentieth century is uncertainty, unpredictability, and skepticism. I said that you can see this almost everywhere. The most important exception is business, where the world view hasn't really changed at all.

It is a short hop from Einstein's 1905 paper to realizing that everything is relative, depending on the observer's position and bias. Everyone is biased. No one is objective. Absolute reality is an illusion.

Gödel's incompleteness theorem

The relativist *gestalt* was powerfully reinforced in 1931 by Kurt **Gödel's incompleteness theorem**, one of the twentieth century's most sublime and devastating pieces of logic.

Gödel may quite possibly have been the most eccentric of top twentieth-century scientists, easily trumping his friend Einstein. After working in Vienna from 1924 on, Gödel fled to Princeton in 1938. The attempt to secure him American citizenship barely survived his long and pedantic exposition of the many grave flaws in the US constitution. He eventually starved himself to death, convinced that his food was being poisoned.

Gödel's incompleteness theorem shattered the dreams of mathematicians by demonstrating that, even in a very simple system like arithmetic, statements could be written down that could neither be proved nor disproved within the rules of that system. Any consistent numerical system generates formulae—to take two simple examples, 'a number is equal to itself,' or 'zero is a number'—that cannot be proved, except by importing axioms from outside the system.

Gödel's proof was not confined to mathematics. Reality, he demonstrated, is a construct, not a given. One implication is that the very process of thinking adds to what we think about … and the process can never be completed. No finite language or system can capture all truth.

So Gödel's theorem really takes the implications of Einstein's theories of relativity one stage further: we can dismiss the possibility of absolute truth. (Incidentally, in 1949 Gödel solved Einstein's general relativity equations in such a way that the entire universe is rotating, and time travel entirely feasible. It is unlikely, however, that Gödel's solution is correct, since his universe was not expanding, and ours almost certainly is.)

Relativity in business

Relativity and the absence of absolute truth should be a central tenet in business. Just to mention some of the important relativities:

◆ The customer's perspective will always be different from the supplier's. Often the differences are large. In the two decades when I was a consultant, I never found that the client's view of what his customers wanted was entirely the same as the customers' view. The only antidote is to keep asking the customers what they want and how good you and your main competitors are on each desired dimension. And to listen to what the customers say. And to act on it. This is 80 percent of what good business is about, and very few people do it well.

◆ Your firm's perspective on what customers want will always be different from the customer's perspective (this is the inverse of the first point) and—here is the nub—it will be wrong in proportion to the distance of the executive from the customer front line.

This is a problem, because decisions tend to be taken well away from the customer front line. Decisions are usually taken in the chief executive's suite or the boardroom. The decisions are meant to benefit the customer. Given the distance and the differences in perspective, it is a safe bet that the decisions won't achieve the desired effect. There are two remedies.

One is to move the decisions closer to the customer, ideally to have the decisions made by the people who deal with customers daily, or even —radical heresy—by the customers themselves. I don't mean decisions on issues such as pricing, where the customer's interests conflict with the supplier's. I do mean decisions on new products, on how production and

service are organized, and on everything to do with customer value
except pricing and margin decisions.

The other remedy is to ensure that the decision makers, especially the
chief executive, are in daily contact with customers.

In my experience the first remedy, although radical, is more realistic
than the second. American humorist Dave Barry makes the point tellingly:

> *My theory is that the most hated group in any large company is the cus-*
> *tomers. They don't know about company procedures or anything about what*
> *you do, which drives you crazy!*
>
> *At the same time, your bosses, who are idiots who don't have to talk to*
> *customers, tell you day in and day out that the most important person in*
> *the world is the customer.*[4]

◆ There is no absolute product quality; it is all perception. The customer's view about product quality is unlikely to be the same as yours.
In fact, customers—another heresy—may not care very much about
quality. Even if they do, other things may be confused with quality, or
imputed to quality. A terrific brand, brilliant advertising, stunning service, fast delivery, or plain and simple market leadership—all these
may be confused with quality.

Clearly, some quality differences are so marked as to be indisputable: a
Cartier watch is better quality than a Swatch; a Mont Blanc pen is superior quality to a Bic ballpoint. But when products are more similar, perceptions of quality may be more important than objective differences.
Are McDonald's burgers really better quality than those from Burger
King? Is a Coke better quality than a Pepsi? Is an Agatha Christie murder story better quality than one from Ngaio Marsh?

Remember that quality is a means to an end, not an end in itself.
Quality is what the customer likes. And if your customers don't value
quality, spend your effort on something else that they like more.

◆ Your main competitor's perspective will not be the same as your own.
Don't imagine that your rival will think like you or even interpret the

same data the way that you do. Try to get inside his head. If this is impossible, just observe what he does and draw inferences about how his world view is different from yours.

The medium is the message

A final, important 'relativist' insight comes from Marshall McLuhan, English professor turned media guru. In his path-breaking 1964 book, *Understanding Media: the Extensions of Man*, he writes:

> *In a culture like ours, long accustomed to splitting and dividing things as means of control, it is sometimes a bit of a shock to be reminded that, in operational and practical fact, the medium is the message ... the personal and social consequences of any medium—that is, of any extension of ourselves—result from the new scale this is introduced into our affairs by each extension of ourselves, or by any new technology.*[5]

What McLuhan means is that media are not neutral: they have their own 'message' and effects, quite independent of content, and often more important than the specific content. The invention of writing facilitated the widespread development of rational and analytical thought. The printing press and printed books—a new medium—had profound consequences. People did not have to go to church to read the Bible or rely on a priest to interpret it for them; they could read it themselves and develop their own interpretation. It is therefore not fanciful to link Protestantism and individualism together with a change in the 'medium'—the advent of mass-produced, cheaper books. The medium was more important than the message (the Bible, for example, which of course had not changed).

When television arrived it had an enormous social impact, quite independent of the content of broadcasting. The medium itself elevated visual impact, downgraded thought, and collapsed time and space by bringing centuries of history and news from around the world into everyone's living room. TV is a 'cool' medium, not in the modern sense, but in that it cools down its content, taking the edge off bloody conflict and splicing it with sexy Levi commercials. TV is (or was) a one-way medium that gives enormous power to image makers and broadcasters and arguably (my opinion rather than McLuhan's) reversed a century of progress in widening and deepening the intellectual powers of ordinary people. The content can be excellent and intellectual, but the passive nature of traditional television viewing, and the inherent preference given to image and emotion over substance and reason, corrode thought and creativity: the medium is the message.

The internet as new medium and new message

What is great about media, however, is that they don't stand still. Cheap telecommunications, a large fax network, and above all the internet are going to transform society at least as profoundly as television, but much more constructively. The internet is a rich medium that connects individuals to other individuals, businesses to each other, and individuals to businesses.

Unlike TV, which was biased toward the 'center' of society and was very much *de haut en bas*, the internet shifts power to individuals and consumers, away from governments, image makers, élites, and big business. The internet favors individual thought and action. It gives power to entrepreneurs and 'insurgent' businesses, and takes power away from established corporations with physical assets and 'legacy mindsets.'[6]

Previous media revolutions had a significant impact on business—television promoted mass marketing and increased the advantages of global scale in consumer goods—but even more impact on society as a whole. The internet will have a terrific effect on society, but may well change business to an even greater extent. Here the medium really is the message. For those in business, it is wrong to think about the internet primarily as a channel of distribution; it will change the nature of business and the specifics of competition in any business in a much more radical way. Those who grasp the enormity of the change, and that the medium is the message, will have an enormous advantage over those who don't.[7]

Summary

The theory of relativity tells us that time is not an independent variable in business. It is part of what we offer. Time is product. Time is service. Time is money. Time is competitive advantage.

Time is also abundant. Organizations and executives waste tremendous amounts of time. When they really want to add value, they do so very quickly and efficiently. Organizational life normally consists of a few islands of high customer value surrounded by oceans of corporate bilge. The typical firm could do things that matter to customers two to three times faster.

Relativity tells us that objective reality is a mirage. We make our own reality. The future is not the past run backwards. The future is what we make it. There are many potential futures. There are multiple routes to success. The raw material of success is all around us, lying unused.

The medium is the message. Business and society are profoundly affected by major revolutions in technology, particularly communications technology. The invention of writing was one such revolution. So were the printing press and the rise of the book industry. So was television. So is the internet.

These revolutions transform our senses. They alter our relationship to space and time. They change the way we interact with each other, the sort of people we think we are, and the sort of things we think we should do. They shift the balance of power in society.

The internet is changing the nature of business reality. The internet makes it imperative that we think in terms of 'product-time' and 'service time,' and that we find new ways to deliver far more rich value, to far more customers, at far less cost, far more quickly. The old absolutes of business are gone. There is a brave new world to fear and to create.

Action implications

◆ *Think service-time.* Dramatically reduce the time you take to deliver products and services to your customers. Think not of products and time, but product-time. Not service and time, but service-time.

◆ *Realize that customers' and competitors' perspectives will always be different from your own.* Struggle to understand their viewpoints. Influence the customer's view by being there, molding customer perception even as it is being formed. Immerse yourself in your market. Don't understand it. Define it. Change it. Create it. Live it.

◆ *Grasp some of the ways in which the new medium of the internet could change the 'message'*—the commercial reality—for your industry, your business, and your career. Find ways to deliver far more rich and individualized value to far more customers at far lower cost and far greater speed. If you can't, then found a new business or find a new career.

Notes

1 My account has borrowed extensively from the very useful essay on Einstein in John Simmons (1996) *The 100 Most Influential Scientists*, Carol Publishing Group, New York. Einstein is ranked number 2, behind Isaac Newton, and ahead of Neils Bohr (number 3) and Charles Darwin (number 4).

2 Mark F Blaxill and Thomas M Hout (1987) 'Make decisions like a fighter pilot,' in Carl W Stern and George Stalk Jr. (1998) *Perspectives on Strategy*, John Wiley, New York, p 165.

3 I have written on this at more length in the chapter on 'Time Revolution' in Richard Koch (1998, revised edition) *The 80/20 Principle: the Secret of Achieving More with Less*, Nicholas Brealey, London.

4 *Fortune*, July 7, 1997.

5 Marshall McLuhan (1964, 1965) *Understanding Media: the Extensions of Man*, revised edition, McGraw-Hill, New York.

6 The phrase 'legacy mindset' comes from the Boston Consulting Group. See the excellent new book by Philip Evans and Thomas S Wurster (2000) *Blown to Bits: How the New Economics of Information Transforms Strategy*, Harvard Business School Press, Boston. Evans and Wurster comment: 'a greater

vulnerability than legacy assets is a legacy mindset. It may be easy to grasp this point intellectually, but it is profoundly different in practice. Managers must put aside the presuppositions of the old competitive world and compete according to totally new rules of engagement. They must make decisions at a different speed, long before the numbers are in place ... They must acquire totally new technical and entrepreneurial skills, quite different from what made their organization (and them personally) so successful. They must manage for maximal opportunity, not minimum risk. They must devolve decision making, install different reward structures, and perhaps even devise different ownership structures.' Good luck!

7   The ways in which the internet will transform business lie beyond the scope of this book, although some useful hints are given in Chapter 11. There are three books that you must read to understand what is happening: one is the BCG book quoted above. The other two are Alex Birch, Philipp Gerbert, and Dirk Scheider (2000) *The Age of E-tail*, Capstone, Oxford; and Evan I Schwartz (1999) *Digital Darwinism: Seven Breakthrough Business Strategies for Surviving in the Cutthroat Web Economy*, Broadway/Penguin, New York/Harmondsworth.

8

On Quantum Mechanics

Quantum mechanics surpasses any other conceptual breakthrough in the breadth of its scientific ramifications, and in the jolt its counterintuitive consequences gave to our view of nature; the micro-world is fully as strange as the cosmos.

Professor Sir Martin Rees, the UK's Astronomer Royal

The opposite of a trivial truth is plainful false. The opposite of a great truth is also true.

Niels Bohr

The most majestic triumph of twentieth-century science

Relativity theory is odd enough, yet it gets weirder. Take a deep breath.

Quantum Mechanics (also known as quantum physics or quantum theory) is the jewel in the crown of twentieth-century science, the most majestic triumph of insight and intellect. Quantum theory shows how the universe, at its most fundamental level, *really* works. Yet it is not a comfortable ride. Even Einstein found quantum theory so strange and subversive that he refused to accept it all, comparing it to 'the system of

delusions of an exceedingly intelligent paranoiac;' hence his famous—
and wrong—observation that 'God does not play dice with the universe.'

So what is quantum physics, and how was it discovered?

Niels Bohr and the quantum leap

The great Danish physicist Niels Bohr (1885–1962) first realized, around
1912–13, that Newtonian mechanics could not explain how atoms
behave. Ernest Rutherford (1871–1937) had already developed a model
of the atom as a miniature solar system with a tiny nucleus of protons and
neutrons orbited by even smaller electrons. It was already known that
atoms were unstable. Bohr guessed that electrons changed orbit when
they radiated light; he therefore identified the emission of a 'quantum'
with the 'jump' of an electron from one orbit to another.

According to Bohr's model, electrons 'excited' by a bombardment of
energy may leap from one orbit to another (inner or outer) orbit,
instantly passing from one position to another, non-adjacent position,
without passing physically in between.

This led to the hypothesis that when an atom has a choice of states to
leap into, it decides entirely at random. A quantum leap (or jump, the
term physicists prefer) is the *smallest* change that can be made, and it happens unpredictably.

Over the next two decades, mathematical modeling of the atom led
to new and counterintuitive insights. Two of the most important of these
are **Heisenberg's uncertainty principle**, and **Bohr's principle of
complementarity**.

Heisenberg's uncertainty principle

In 1927, Werner Heisenberg proved that uncertainty is inherent in the
equations of quantum mechanics. He showed that if we try to measure
both the position and the momentum of an electron, we will fail. The
more accurately we know where the object is, the less certain we can be
of its momentum; and the other way round. The uncertainty principle
therefore states that it is not possible to calculate accurately both the posi-

tion and the momentum of a subatomic particle. Language cannot effectively describe the atom; all we can do is measure the atom, but only with inherent uncertainty. This is not because of any defect in our measuring techniques; it is because of a fundamental fuzziness in how tiny matter behaves. We will never know whether atoms do or do not behave like precise points moving at precise speeds.

Heisenberg therefore slammed the assumption that 'behind the statistical universe of perception there lies hidden a 'real' world ruled by causality' as 'useless and meaningless.'

The principle of complementarity: wave–particle duality

In 1927, Bohr expounded his principle that light is both wave-like *and* particle-like, simultaneously. The reality detected depends on the experimenter and his method. If you observe a photon with a particle detector, you get a particle. Observe the photon with a wave detector, and, hey presto!—you get a wave. Neither, Bohr said, is more real or accurate. We can only see a wave or a particle at any one time, yet both descriptions are necessary to get a full sense of what light is. The two methods complement each other.

'Both/and' becomes a better perspective than 'either/or.'

Schrödinger's cat

Erwin Schrödinger (1887–1961) was a colorful Austrian physicist who in 1925 created the wave equation, describing the electron's behavior around the atomic nucleus. Bohr told Schrödinger:

> *Your wave mechanics has contributed so much to mathematical clarity and simplicity that it represents gigantic progress over all previous forms of quantum mechanics.*

Nevertheless, Schrödinger, like Einstein, was upset by the failure to find an underlying cause of atomic and subatomic behavior. Einstein and

Schrödinger both wanted to detect the underlying reality that they supposed must exist below the unreal world of the quantum. To demonstrate the absurdity of quantum theory's probabilistic conclusions, Schrödinger created the parable of the cat that is both dead and alive.

His thought experiment goes like this. Imagine a live cat in a black box, which we can't see into. The box also contains a phial of poison that, if released, will kill the cat. There is a 50/50 chance that radioactive decay will release the poison. Common sense suggests that, at any moment, the cat is either dead or alive; either the poison has been released or it hasn't. The fact that we don't know whether the cat is dead or alive is irrelevant; it's either quite happy or quite dead. Yet quantum theory says that the radioactive material does not 'decide' whether to decay or not until it is observed. Until we open the box, therefore, according to quantum theory, the atomic decay has neither happened nor not happened. The cat, in turn, therefore, can neither be dead nor alive; it must be *both dead and alive*. Only by opening the box do we make it dead or alive.

Pretty silly? Schrödinger thought so. What he highlighted was the gap between quantum reality—the world of very small pieces of matter—and reality beyond the micro-world; that is, the everyday reality of cats and people and anything bigger than an atom. But although he would have liked to, Schrödinger did not disprove the nature of quantum reality. Experiments have proved beyond any doubt that quantum mechanics is correct in saying that photon particles behave randomly and yet in a related manner. One photon has an instantaneous effect on another proton, even when logically it can't. The fundamental particles that make up the world seem to be inseparably connected to each other, part of some indivisible whole; the particles 'know' what the others are doing.

What I think Schrödinger's parable of the cat does succeed in doing is to question the relevance of quantum theory to phenomena outside the quantum world, and to inject some cautious common sense into any attempt to extrapolate from the micro world.

How relevant is quantum theory to the non-micro world?

Quantum mechanics has many puzzling paradoxes, but one that is of particular interest is the contrast between the impressive, practical results of the theory on the one hand, and its mysterious, spooky and apparently illogical nature on the other hand. No one would believe quantum theory if its scientific predictions had not been validated so many times, or if it had not changed our lives so powerfully. Quantum theory can predict experimental results to many decimal places. Quantum theory (and relativity) have given us the transistor, the microchip, nuclear power, and lasers. Modern cosmology would be impossible without quantum theory.

Here we have this amazingly successful physical theory, and yet the implications of quantum theory—even within science, and within the science of microscopic pieces of matter—appear very woolly, peculiar and wrong-headed, even to eminent physicists such as Einstein and Schrödinger. So the question we must now pose is: how relevant is quantum mechanics to anything else, and specifically to business?

Here we must wade into the choppy waters of controversy. Various far-reaching interpretations of quantum theory have been given by writers on 'pop' science, some of whom claim that there is a new 'quantum' theory of society and business that replaces the allegedly dominant 'Newtonian' theories. I find these claims unconvincing, for reasons I'll cover later in this chapter. Before outlining a good example of quantum theory extrapolated to business, and challenging this interpretation, I would like first to draw out two positive points where I think it *is* valid and instructive to use quantum theory as a metaphor for business purposes.

Escaping the tyranny of either/or

Both Heisenberg's uncertainty principle and Bohr's principle of complementarity are direct exponents of something that was at the time, in the 1920s, new in science: a both/and rather than an either/or approach. The methodology was a response to the nature of quantum reality. Whereas it would make no sense to say, for example, that a cat was both in the drawing room and in the bedroom simultaneously, it *does* make some sort of sense, according to Heisenberg, to say that a particle could be in two places at once; at least, given the existence of quantum leaps and discontinuities within atoms, we cannot visualize accurately their position and the momentum. And, Bohr adds, light can both be a series of waves and a stream of particles, at the same time. Both descriptions, Bohr says, are necessary and complementary, to help us understand what is going on.

The discovery of both/and as a sensible alternative to either/or is, I believe, a valid and relatively novel way of thinking that can be derived, at least loosely, from quantum mechanics. Of course, if quantum mechanics had never existed, we could still have made the mental breakthrough to both/and thinking.

For one thing, we could have deduced both/and thinking from modern business itself. Name almost any trade-off that used to be regarded as an either/or issue, and it can be shown that, at least in some circumstances, the trade-off can be avoided and a both/and solution created. For example, if we are sufficiently creative or lucky, we can *both* keep the stock market happy *and* be socially responsible. We can *both* aim for high profits *and* deliver top value to customers. We can have *both* high quality *and* low costs; quality may not just be 'free,' but actually have 'negative costs.' You can gain from a particular transaction, *and* so can I. In the economy at large, we can now apparently even have high growth *and* low unemployment; we can also enjoy high growth *and* low inflation; and low unemployment *and* low inflation. Previously 'inevitable' trade-offs can now—sometimes—be transcended. Whenever a trade-off is proposed, we should assume that there is a way round it, so that we can have our cake and eat it. There probably is, if we are creative enough.

Testing out multiple possible options

A second useful, although admittedly weaker, point is an analogy between the behavior of electrons (or photons) and the value of testing out multiple possible options. Here I would like to quote Danah Zohar and Ian Marshall, writers of popular books on quantum theory and its social implications:

The atom may become unstable for no apparent reason ... So, quite suddenly, the electrons in a previously stable atom may begin to move into different energy orbits ... there is no way of knowing by which path a particular electron may travel ... Indeterminacy—the lack of any physical basis for predicting the outcome of events—characterizes the quantum realm. The electron may go to the next highest state, it may leap over several intermediate states or even double back on itself ... quantum physics tells us that the electron actually follows all these possible paths, all at the same time. It behaves as though it is smeared out all over time and space and is everywhere at once.

In much the same way as we play with multiple possibilities in our imaginations, or launch 'trial balloons' to see how something might work out, the electron puts out 'feelers' ... to see which path ultimately suits it best.[1]

The business analogy is that we should conduct thought experiments, and actual experiments, before settling on an important course of action. Imagine doing *a*, *b*, and *c*; and also doing *x*, *y*, and *z*. Neither you nor your firm is predestined to pursue one course of action. You have very many degrees of freedom, bounded only by your imagination. There is *always* another path, always another way of doing things, always some place else to go. Create many possibilities. Play with them in your mind. Consult colleagues. Think the unthought. Think the unthinkable. Narrow the choice, but don't fix yet on one course of action. Keep a few in your locker.

Now put out feelers to collaborators, and to the marketplace. Ask distributors, suppliers, customers, trusted experts. What do you think about *x*? Would you like *y*? What would be the reaction if we *w* + *z* but also − *b*? Even now, do not decide what you are going to do.

Launch test markets to see whether the concept, technology, and product variants can meet the product concept expectations, and whether the product is sustainable: whether there is a sufficiently high level of repurchase. See which trial balloons go up and stay up, which go down, which go nowhere. Experiment without commitment. Now revise your plans. Conduct more experiments. Iterate.

When you are ready, go. But not until you have behaved as if you were an electron.

Quantum theory used as a battering ram

I think that these two examples of management insight from quantum mechanics—the both/and point and the idea of putting out feelers before moving decisively—are reasonable and helpful analogies for business from the micro-world. But several writers have gone further, and attempted to use quantum mechanics as a battering ram to attack the 'Newtonian' world view and support what we may loosely term a 'new age' philosophy of society and business, based allegedly on quantum physics. This 'new age' use of quantum theory is interesting, bold and coherent. It is also wrong.

After describing the new view, there are three issues that I would like to pursue. One is whether the 'new age' view can derive legitimacy from quantum physics. The second issue is whether, regardless of its provenance, the new view deserves the attention of businesspeople—whether it contains helpful insights. A final question is whether the new view is really consistent with the most important positive insight that can legitimately be drawn from quantum mechanics.

First, the description. For many years, the leading exponent of the 'new age' view of quantum physics (although I think she would repudiate the new age tag) has been Danah Zohar, a writer who has studied physics, philosophy and religion. Her views are clearly expressed in three intriguing books, *The Quantum Self*, *The Quantum Society*, and *ReWiring the Corporate Brain*. In *The Quantum Society*, she writes:

> *I believe we must come to appreciate that self, society and nature all derive from a common source, that each is a necessary partner in some larger creative dialogue ... there is just one reality, and we are part of it. Physics tells us about the processes of creativity and transformation in the natural world ... If we understand the actual physical basis of transformation, perhaps we can align ourselves with it ... Quantum physics in particular almost cries*

out for use as a more general model for a whole new kind of thinking about ourselves and our experience. There is an uncanny and intriguing similarity between the way that quantum systems relate and behave and so much that we are now beginning to understand or hope for about human social relations.

In *ReWiring the Corporate Brain*, Zohar describes 'the choice between Newtonian management and quantum leadership.'

In a 'Management–Leadership Chart,' she says:

Newtonian Management Stresses Certainty, Predictability, Hierarchy, Division of labor or function fragmentation, Power emanates from top or center, Employees are passive units of production, Single viewpoint; one best way, Competition, Inflexible structures; heavy on bureaucratic control, Efficiency, [and] Top Down (reactive) operation.

By way of contrast:

Quantum Management Stresses Uncertainty, Rapid change; unpredictability; Nonhierarchical networks, Multifunctional and holistic (integrated) effort, Power emanates from many interacting centers, Employees are cocreative partners, Many viewpoints; many ways of getting things done, Cooperation, Responsive and flexible structures; hands-off supervision; Meaningful service and relationships; [and] Bottom-up (experimental) operation.[2]

Does quantum theory sustain a new view of management?

Let us leave aside, for the moment, the question of whether 'quantum management' is better than 'Newtonian management.' The first issue is whether 'quantum management' can reasonably be derived from quantum theory.

I have allowed above that it is reasonable to derive from quantum theory, by way of analogy, two of the eleven points that Zohar makes about 'quantum management,' namely what we have called both/and, which she calls here many viewpoints, and the point about experiments. I can also see that analogies with the behavior of the micro-world can just about sustain the points about uncertainty, and rapid change, and unpredictability (although many other observations from the natural world would make the point about uncertainty equally well, and while it is clear that the micro-world is, to us, unpredictable, I am not sure that it typically demonstrates rapid change). Let us, however, give the benefit of the doubt here.

But is there anything at all in quantum physics that demonstrates non-hierarchical networks, or multifunctional and holistic (integrated) effort, or that power emanates from many interacting centers, or that employees are cocreative partners, or cooperation, or responsible and flexible structures, hands-off supervision or meaningful service and relationships, or indeed the bottom-up part of bottom-up (experimental) operation? It may be my lack of expertise in quantum mechanics that leads to my puzzlement, but having read many accounts of quantum theory I can see no shred of evidence that it supports these latter points about 'quantum management' in any way whatsoever.

No doubt Zohar would claim that the important thing is not the particulars but the general *gestalt* of quantum physics, and other twentieth-century science, as opposed to Newtonian physics. And here there is *something* of a point. Newtonian physics *did* crystallize and popularize the mechanical view of the world, the idea that the world and its parts were a series of machines that behaved predictably and in an orderly fashion. Quantum physics, in partial contrast, reveals that *some parts* of reality, namely submicroscopic and inanimate particles and their ilk, behave in a way that is difficult for us to predict and that certainly cannot be characterized as machine-like. Is this a sufficient basis for a new world view? And can it then be reasonably extrapolated to give a new view of management? Not, in my view, without a lot of help and bucketloads of imagination.

Some help, it is true, comes from other twentieth-century science: from some that we have looked at already—relativity, and Gödel's incompleteness theorem—and some that we shall examine in Part Three, namely the concepts of chaos, complexity, and non-linear systems generally. Together these do reveal that there are important parts of the natural world that behave in interesting patterns that are not linear or mechanical, and are often difficult to understand at all. We have also seen that nineteenth- and twentieth-century biology supports the importance of cooperation (mentioned by Zohar) and variation (not given prominence by her). But is it fair to put all this together in the Zoharian cocktail of 'quantum management' (which is implicitly good) and contrast it all to 'Newtonian management' (implicitly bad)? I would say at the very least that the case is not proven. I would also make the very important point that quantum physics does not *replace* Newtonian physics, which remains valid, or only very slightly inaccurate, for everything that it always covered. Does this imply that 'Newtonian management' is almost perfectly valid for large parts of our experience too?

Is 'quantum management' better than 'Newtonian management'?

The second issue we should briefly consider is whether 'quantum management' is a useful business philosophy, regardless of whether it does or does not fairly derive from quantum physics. 'Quantum management' is a mixed bag of attributes, some of which are certainly useful in particular circumstances. But it is interesting that there is no example on earth of a business organization that is run more on 'quantum management' than 'Newtonian management' lines. Of course, this could just be timidity and unfamiliarity with the new paradigm. But it could equally well be because the task is impossible.

Is 'quantum management' really compatible with valid insights from quantum theory?

This is our third and final issue. For me, the key insight that can directly be drawn from quantum mechanics is that, wherever possible, we should adopt a both/and approach to business, escaping the tyranny of either/or. And here it is deeply ironic that expositions of quantum thinking, like those of Danah Zohar, should take such pains to denigrate Newtonian, mechanistic thinking. The paradoxical implication is that we can have *either* 'Newtonian management' *or* 'quantum management.' This is doubly inappropriate. The choice we have is not between Newtonian physics and quantum physics: we need them both. And we are not forced to choose between the Newtonian world view, or a so-called Newtonian view of management, and the perspectives given by quantum theory and other twentieth-century science. We can benefit from both.

The appropriate view, surely, is that quantum theory reveals another world, a parallel universe that we cannot directly experience, another set of insights and language; a world from which we can draw insight and inspiration, but which cannot and should not replace the Newtonian world of rational thought, plans, objectives, solid engineering, numbers, budgets, and the quest for high returns. Quantum thinking has a valuable role in supplementing this approach, in drawing attention to its flaws, and in stimulating the imagination. But quantum thinking alone can't supply a way of managing a business organization, any more than it can build a bridge or send someone to Mars. We do quantum physics and quantum thinking no favors by exaggerating its reach or comprehensiveness, or by setting it up as a complete replacement for 'Newtonian' methods. The insistence by some writers that the quantum way is the true way, and the extrapolation of quantum reality to cover all reality, runs directly counter to the more tolerant and inclusive wisdom of Niels Bohr, probably the greatest quantum physicist ever, who insisted that 'the opposite of a great truth is also true.'

The quantum world is not better or worse than the Newtonian world; it is just different. *Vive la différence!*

Summary

Quantum mechanics shows how the smallest particles of matter behave: the answer is 'oddly, unpredictably, bizarrely.' Werner Heisenberg showed that uncertainty is inherent in quantum mechanics: we cannot follow the position and momentum of electrons. They seem to move in different directions at random and even simultaneously. Niels Bohr showed that light is, simultaneously, both like a wave and like a particle. Both descriptions are necessary and complementary for us to grasp what is happening in the micro-world. Erwin Schrödinger's parable of the cat shows the gulf that separates quantum theory's indeterminate conclusions from the reality of ordinary life beyond the micro-world.

We must be cautious in extrapolating from quantum theory. The mysterious world of microscopic, inanimate particles adds a new perspective on life, but it does not invalidate other, more straightforward perspectives. Even in physics, quantum mechanics supplements, but does not replace, Newtonian mechanics. Quantum theory supplies extremely minor adjustments to Newtonian calculations, adjustments that for nearly all practical applications can be safely ignored. Quantum theory is a marvelous triumph of the intellect, and has led to computers, biotechnology, and genetic engineering; yet its impact on mainstream engineering has been slight.

Attempts to use quantum theory as a battering ram to attack 'Newtonian,' 'mechanical' attitudes and ways of behavior are unconvincing. Such attempts betray a shallow understanding of the nature of science and business alike, and would probably have been repudiated by the 'fathers' of quantum theory itself.

Two important insights for business may, however, be reasonably drawn from quantum theory. If light can be visualized both as a series of waves and as a stream of particles, then any one view of business activity is likely to be restrictive. For example, business can legitimately be viewed as an economic exercise to maximize profits and cash; but this does not preclude the simultaneous legitimacy of an alternative view, for example that business is a creative social activity that exists to add value to customers. These perspectives appear to be incompatible, but they may actually be complementary. By extension, and using a little poetic license, we could generalize that quantum theory implies that the world is not either/or—it is both/and. Contradictions can be transcended. With enough imagination, no trade-offs may be inevitable.

The second valid insight is the value of testing out possible multiple options. As far as we can tell, electrons can and do go anywhere and everywhere all at once. We can too, via thought experiments. These can be followed by putting out feelers to the market and launching some trial balloons. A useful rule of thumb may therefore be: Don't commit major resources until you've lived like an electron.

Action implications

◆ *Pursue both/and.* Don't believe that contradictions and trade-offs are inevitable. Be creative in seeking ways to defuse and defeat them, so that you can deliver the best of both worlds. Escape from the limitations of 'either/or' thinking. Have it all.

◆ *Experiment in thought and deed before you commit yourself to a major initiative.*

Notes

1 Danah Zohar and Ian Marshall (1993) *The Quantum Society*, Bloomsbury, London.

2 Danah Zohar (1997) *ReWiring the Corporate Brain*, Berrett-Koehler, San Francisco, p 87.

Part Two Concluding Note

In Part Two, we've examined two different types of physics, two different world views, and, by implication, two different views of organizations and management. It's odd that the same physical science can give rise, sequentially, to two such different views of how the world works. And, on reflection, the new world of quantum physics, with its fresh and subversive implications for how the universe works, supplements but does not replace its older cousin, the physics derived from Isaac Newton.

Chapter 6 celebrated the Newtonian model, and the power of numbers, analysis, and simple ratios comparing two key relationships. Here lie the origins of accounting systems, profitability analyses, and two-by-two matrices. Their power is not to be sneered at. We examined one such powerful relationship: that between profitability and the distance from competitors and their relative size. While the extent to which the relationship works mathematically is unproven, we explored the intuitive appeal of the concept of the 'gravity of competition,' where competition from near and similar business units depresses returns. We posited that the way to earn very high returns on capital was to establish the maximum possible distance from competitors, along the dimensions of customer type, product type, and geography.

Chapter 7's review of relativity concluded that time should not be thought of as a separate dimension of business. Time must be a key, integral part of any business offering. We also saw that objective reality is a mirage. If reality is a construct without objective underpinning, the way we think, our business endeavors, and our markets may determine our success or failure just as much as the other attributes we bring to bear.

Chapter 8's fascinating journey into the micro-world revealed how strange and difficult to understand are the most minute portions of matter.

In concluding our review of physics, we have refused to let quantum mechanics overshadow Newtonian mechanics. The new does not replace the old. Apart from the world of very small—subatomic—matter, Newton's theories still work. And in business, it's quite possible to be very successful using Newtonian tools, and disregarding everything else.

However, the most important lesson from quantum theory is that apparently irreconcilable positions may just be two sides of the truth: that we should accept great truths as complementary even when they appear to be opposite. This means that in business, as in physics, there is room for the old view and the new one: for rational truth, numbers, cause-and-effect reasoning, and the mechanical view of the world; and also for unpredictable, probabilistic, experimental, unfolding, and paradoxical patterns of reality—about which we shall learn a great deal more as we look into the world of non-linear systems in nature and society in Part Three.

Part Three

The Non-Linear Laws

Interdisciplinary Science

Introduction to Part Three

In Part Three, we turn to non-linear systems and to the interdisciplinary science that addresses them: to chaos, complexity, and systems thinking generally.

According to non-linear science, the world does not behave in a linear fashion, with Newtonian regularity, with cause and effect discrete and identifiable. The mathematician Stanislaw Ulam commented that calling the study of chaos 'non-linear science' is like calling zoology 'the study of non-elephant animals.' Ulam has a point, because the modern world is full of non-linearity, but he was exaggerating.

It should be stressed that some things *are* linear, and here Newtonian physics still work. But many of the important and interesting things in life are non-linear systems. The whole is more than the parts, and comprises something qualitatively different from the parts.

Linear analysis can't help when studying complex systems. But the relatively new interdisciplinary sciences of **Chaos and Complexity** can. We review these in Chapter 9.

In Chapter 10, we find out how to achieve more with less, with the tool of the **80/20 Principle**. We also derive insight into non-linear systems from the 'power of weak ties,' Von Foerster's theorem (from cybernetics), control theory, and Fermat's principle of least time.

Chapter 11 covers the non-linear aspects of growth, technological change, networks, and the 'new economy.' It revisits **Punctuated Equilibrium** and introduces the related and extremely useful concept of the **Tipping Point**, which helps us predict whether and when a new trend or product is going to take off. It shows the key role of technology in driving growth. The chapter also introduces the **Law of Increasing Returns**, which overturns most of traditional microeconomics and shows how the construction of temporary monopolies may be both inevitable and in the public interest.

Chapter 12 looks at how systems theory can help us sustain success, despite the traps of the **Paradox of Enrichment**, the **Law of Entropy**, and the **Law of Unintended Consequences**.

9

On Chaos and Complexity

Chaos is when any system is so complex and irregular that it appears to be random unless you know a lot of hidden information about it. Chaos is lovely, it is absolutely wonderful. It is full of all sorts of intriguing forms and behaviours.

Ian Stewart

The third great scientific breakthrough

Chaos and **Complexity** are probably the most important scientific innovations of the late twentieth century. They are interdisciplinary concepts and fields of study, drawing on mathematics, biology, physics, economics and many other disciplines. Although it is too early to be sure of their place in the history of science, many believe that they rank alongside or just behind relativity and quantum theory, comprising the third great scientific breakthrough of the last century. Chaos and complexity also throw up insights that are very congruent with, but also complementary to, those already provided by relativity and quantum theory. As one physicist comments:

Relativity eliminated the Newtonian illusion of absolute time and space; quantum theory eliminated the Newtonian dream of a controllable

measurement process; and chaos eliminates the Laplacian fantasy of deter-
ministic predictability.[1]

Chaos

As suits its Alice-in-Wonderland nature ('whenever I use a word it means
what I want it to mean'), 'chaos' in this context means 'the concept or field
of study of chaos.' Very few scientists working on chaos call it 'chaos theory;' and 'chaos' is not necessarily or even usually 'chaotic' in the ordinary
sense of the word. Let's face it, chaos is a bad name, because the processes
studied often reveal beautiful and intricate patterns. The processes are only
superficially chaotic; underneath there is a deep, if irregular, order.

Chaos is the search to identify and understand non-linear patterns that
have been ignored by traditional science. The fun of chaos is that if you
look at structures in non-linear systems they are very similar, regardless of
the phenomenon: it could be the weather, the economy, the way cities
are organized, a set of numbers, snowflakes, coastlines, stars in the sky, or
a series of stock exchange prices.

Chaos has its roots in quantum theory and in mathematical work, during the nineteenth and early twentieth centuries, on chance and probability. Quantum theory tells us that very small things like atoms and
photons do not behave in a linear or predictable way. They really are
chaotic. The brilliant French mathematician Henri Poincaré wrote with
great insight in 1908:

> *A very small cause, which escapes us, determines a considerable effect which*
> *we cannot ignore, and then we say that this effect is due to chance.*[2]

Sensitive dependence on initial conditions

Poincaré is really the intellectual forerunner of chaos, where the central
insight is **Sensitive dependence on initial conditions**. Many physical systems exhibit sensitive dependence on arbitrary initial conditions, and are
therefore essentially unpredictable. The classic example is the weather. It

used to be thought that, with enough data and computing power, it would be possible to predict the weather reliably, months in advance. We now know that this is impossible, because weather is subject to extremely sensitive dependence on initial conditions.

The butterfly effect

In 1972 Edward Lorenz, a meteorologist at the Massachusetts Institute of Technology, gave a paper provocatively entitled 'Predictability: does the flap of a butterfly's wings in Brazil set off a tornado in Texas?'[3]

The question, he said, was unanswerable but illustrates the nature of the weather. For many years Lorenz had used computers to model the weather, hoping to improve long-term forecasts. His pioneering work showed that eddies and cyclones obeyed certain mathematical rules, yet never repeated themselves. Long-term weather forecasting, he concluded, was impossible, for extremely interesting reasons. Although he could model the influences on weather, minute changes in a couple of variables, extrapolated over a month or beyond, could produce totally different results.

Lorenz's insight was not just that small effects can have huge consequences—this is an old story, embodied for example in the Prussian verses about a kingdom being lost for want of a horseshoe nail. His real breakthrough was to demonstrate that the future of weather was literally uncertain, even if we knew everything there was to know about all the influences on it. The weather makes itself up as it goes along, as does evolution, and as do vibrant economies.

The search was on to understand complex, non-linear systems.

Most interesting things in the universe are complex systems—and 'chaotic'

Both chaos and complexity, like quantum physics, are based around the realization that many things in the world are not linear, not easily predicted, and not simply the sum of their parts.

Complex systems—like the weather, cities, economies, galaxies, insect colonies, packs of wolves, brains, and the internet—are not stable and spend little or no time in equilibrium. Yet they can all be described, and often analyzed, by using the concept of chaos (and, as we shall see later in this chapter, the related concept of complexity). Many of these systems return repeatedly to a position close to where they were before.

Complex systems may have simple causes. Much of their behavior may be described by simple equations: for example, the way that leaves are blown by the wind. And chaos has shown that simple rules of behavior can lead to amazingly complex results, but, again, ones that can be understood in terms of a set of simple subsystems. Certain characteristic patterns recur, yet with infinite, unpredictable variety.

Insights from chaos into the physical world

One milestone in the emergence—and naming—of chaos was the work of mathematician James Yorke around 1970. Yorke and other mathematicians pointed out that when confronted with non-linear systems, the typical response of mathematicians was to attempt to solve them by substituting linear approximations. Yorke demonstrated that this was unnecessary: even non-linear systems that were very sensitive to initial conditions could actually be modeled. Computers could take, for example, biological data relating to fish populations, and produce a graph. Regularities could then be detected that were totally counterintuitive.

What's the connection between cotton prices and the Nile?

At about the same time Benoit Mandelbrot, also a mathematician, was working in IBM's pure research department and using its most powerful new computers to analyze cotton price data. He showed that there were patterns for daily and monthly price changes that matched perfectly; the degree of variation had remained constant over 60 years. There was unexpected order within the disorder. Mandelbrot found the same patterns in all the data he analyzed, including variations in the level of the river Nile over several millennia.

The more research was done, in meteorology, in biology, in geology, in physics, in chemistry, in economics, and in many other areas, the more it became apparent that there were unsuspected regularities that could be described by the relationship of large scales to small scales.

Mitchell Feigenbaum, a physicist at the Los Alamos National Laboratory in New Mexico, soon after conducted a series of calculations to measure the size difference between geometrically converging sets of data, such as Mandelbrot's cotton prices. His calculator repeatedly came up with the same number: 4.669. Feigenbaum had stumbled across one of the most startling properties of chaotic systems: universality. Universality means that, on some dimensions, different systems will behave identically. At a conference at Los Alamos in 1976, one of Feigenbaum's colleagues commented:

> It was a very happy and shocking discovery that there were structures in non-linear systems that are always the same if you look at them the right way.

How long is the coastline of Britain?

Mandelbrot showed that there is no right answer to this question. It actually depends on your measuring instrument. With a small enough ruler, measuring every tiny indentation, the distance approaches infinity!

Fractal similarities

Mandelbrot coined the very useful word 'fractal' to describe things that are very similar to each other, yet not identical—things like coastlines, clouds, cotton prices, earthquakes or trees. Patterns are endlessly repeated, yet also with endless and unpredictable variety. Plotting data from non-linear systems reveals strikingly similar patterns regardless of the actual data being plotted. For example, the year-to-year graph of cotton prices looks eerily like the *shape* of the month-to-month cotton price variations, even if the scales are different.

Business is fractal: no situation is quite like another, but there is a limited set of key factors that always resemble each other. Business outcomes are utterly unpredictable, which is why the quest for a deterministic science of management—if you do x and y, then z will result—is futile and naïve. Yet there are recurrent patterns that are worth studying and recognizing. The fact that business is fractal is the best justification for the case-study method used in business schools, although this would be much more useful if we could map the different fractal patterns for different types of businesses; something no one has yet done.

Research into fluid turbulence by David Ruelle, in mathematical physics in the early 1970s, led to the idea of a strange attractor, a fractal object that is a point and can be modeled mathematically to explain turbulence. Strange attractors could also depict the chaotic behavior of a rotor, with extreme mathematical precision. The theory has since been used in astronomy to explain how stars form 'islands' and 'chains of islands' in the sky, and in fact to map the trajectory of any dynamic system that is sensitive to initial conditions.

Although not strictly linked to chaos, let's digress briefly to look at two laws of chance, one from statistics and probability theory, and one from the study of chance in history.

Principle of impotence

This theory says that it's impossible to devise a successful gambling system to be used against a fair coin or other genuinely random device. Where there is chance involved, it may be impossible for managers to do better than flip a coin. More generally, at what point do you make an important decision? When you are 70 percent sure that it's right? 80 percent? 95 percent? Whatever percentage you choose, remember that your estimate may be way off, and will usually be too optimistic. Remember also that analysis and delay have costs.

Fear not. The **Principle of impotence** provides a good excuse for action based on insufficient data. Of course, you shouldn't decide cavalierly. But equally, don't allow a difficult decision, where there will always be great uncertainty, to paralyze your progress.

The hinge factor in history

Historians have long known that chance events can turn the course of history. The crucial event—the 'hinge'—may be something small, banal and unexpected. A recent book[4] gives many interesting examples of the **Hinge factor in history**. Here are three.

In the American civil war, the Confederate states might have won easily and early had General A P Hill not lost the plan of attack drawn up by his superior, General Robert E Lee. Hill used Lee's handwritten plan to wrap a couple of cigars. These were mislaid and found shortly afterwards in the abandoned Confederate camp by a sergeant from the opposing Union Army, who passed it on to his supreme commander, General McClellan. The battle of Antietam was a surprise Union victory as a result.

A similarly trivial blunder sealed the fate of Germany's *Bismarck*, the world's fastest and most powerful battleship. Other things being equal, the *Bismarck* should have been able to block supplies from America to Britain during the Second World War. But in 1941, Admiral Gunther Lutjens neglected to fill her up with fuel in his haste to leave Norway and sink the British battleship *Hood*. Although the latter went down with 1500 fatalities, the battle caused slight damage to two of the *Bismarck*'s fuel tanks, and this, together with the failure to fill up in Norway, meant that the ship had to travel much slower than normal as it made for Occupied France. British warships were therefore able to catch and sink her.

Why did communism collapse in 1989? Apart from the vanity of Gorbachev, who played footsie with the West when he could equally well have maintained repression, a very important catalyst was an unscripted chance reply during a TV interview with Gunther Schabowski, who'd just been appointed Communist party spokesman in the German Democratic Republic (the old East Germany). 'When,' the interviewer asked, 'would East Germans be allowed to travel freely to the West?' Schabowski shot back petulantly, 'They can go whenever they want, nobody will stop them.' The audience was stunned, then there was bedlam. Thousands of East Germans poured over the border, and communism was dead.

Chaos and complexity—and history—tell us that business is uncertain and risky. Past success gives no exemption. *Ergo*, take risks: Outcomes are uncertain anyway. Evolutionary psychology says that when we are comfortable, we take the fewest risks, despite the fact that we can afford to. Because too few people take risks, the rewards from doing so are, in aggregate, more than they should be (in business; this does not apply to gambling, where the odds are precisely correct when there is no 'house,' and miserable when someone else is making a living out of it). Successful businesses take too few risks. As long as you avoid 'bet the company' moves, take as many risks as you intelligently can.

Chaos, chance, and business

The concept of chaos is that although the world largely comprises non-linear systems, there are patterns discernible within the irregularities. Disorder in the universe is constrained. Chaos and chance do not lend themselves to tracing simple, causal links, which is what turns on most of us in business. Yet there are some very useful morals for business. Here are eight.

There is always some pattern or order in apparently random or disordered data

Patterns exist. The only question is whether we can detect them. All markets generate patterns of behavior and response.

Analysis may not be the best way of finding the hidden order

Analysis may be incapable of finding the answer, if the system is reasonably complex and interdependent. The human brain has the flexibility and imagination to discover the pattern. Therefore, if you want to understand a market, it may be better to immerse yourself in it, and wait for inspiration to come, than to search for data and analyze it.

Simple systems do complicated things

There may be three or four key things that, combined with 'chance' (better called sensitive dependence on initial conditions), lead to incredibly complicated behavior. Try to isolate the key variables, the main causes, that interact with each other. However, resist the temptation to reduce everything to one main cause or effect.

Complex systems can give rise to simple behavior

Behavior is a better and easier guide to a complex system than complex structural analysis. When looking at complex systems such as markets, customers or competitors, look for characteristic patterns of simple behavior. For example, if a competitor always follows your price changes, this is all you need to know; analysis of its decision-making process would be redundant. Always be alert to reliable patterns of simple behavior.

Chance—the role of luck

'Luck' and 'chance' are not necessarily accurate descriptions of unexpected results, but most markets, companies, and business units are complex systems that are sensitive to initial conditions. Therefore expect the unexpected, and expect it often to be due to minute and undetectable causes. There are several corollaries:

◆ Don't expect to be able to control everything. Don't be thrown totally off course when the unexpected happens.
◆ Build flexibility into your plans. If x happens, do y. If w happens, do z.
◆ When something goes wrong, don't waste enormous effort investigating what went wrong and punishing the wrongdoers. Annoyingly, there may be no wrongdoers. Or the 'wrongdoers' may have behaved impeccably, as per your instructions. Get on with working out what to do next.

◆ When something goes right, remember that it may not be at all due to your skill or that of your firm. It may be sensitive dependence on initial conditions that happened to suit you brilliantly. Exploit the trend for all it is worth, but don't believe your own propaganda. The next 'sensitive dependence' may suit a competitor better.

Chance—the need for multiple strategies

Where there is major uncertainty about how an industry may evolve, it may make sense to have more than a single strategy.

Eric D Beinhocker[5] comments that in 1988, when he wondered around Comdex, the computer industry's trade show, there was something very odd and ambivalent about the Microsoft booth:

> While most booths focused on a single blockbuster technology, Microsoft's resembled a Middle Eastern bazaar. In one corner, the company was previewing the second version of ...Windows ...In another, it touted its latest release of DOS. Elsewhere, it was displaying OS/2 ...[and] major new releases of Word, Excel ... [and] SCO Unix...
>
> 'What am I supposed to make of all this?' grumbled a corporate buyer standing next to me. Columnists wrote that Microsoft was adrift ...[and] had no strategy. Reporters told stories of infighting at the company as one group ... worked furiously on Windows and DOS while others poured their energies into OS/2, Mac applications, and Unix.
>
> ...in 1988 it wasn't obvious which operating system would win. In the face of this uncertainty, Microsoft followed the only robust strategy: betting on every horse to win.

Microsoft had *strategies* rather than *a strategy*.

Focus is a wonderful thing, but corporations in fast-moving and unpredictable markets may need to take some of their resources and have a few side bets at long odds: this is equivalent to spending money on financial 'call options.'

As Beinhocker comments:

*A company should use most of its resources to build its current activities,
but the resources devoted to riskier experiments further afield are critically
important, since they could contain the seeds of success in a currently
unimaginable future.*

The first-mover advantage

Since most complex systems are very sensitive to initial conditions, it
makes sense to get in on the ground floor of any new development that
may be important to your key markets. The idea of the **First-mover
advantage**—that the first person in has an edge over other equally qualified later-comers—is well known in business, but the science of chaos
reinforces its importance. A firm that, in an embryonic market, puts out
a product that is, say, 10 percent more attractive than any other offering
may end up with a 100 or 200 percent greater market share, *even if competitors later provide something better.* The first-mover advantage 'locks in'
standards and makes the market behave in ways that are tilted to favor the
first mover.

A cute illustration of this from chaos theorists is the way that clocks
behave. Why should nearly all clocks exhibit 12 hours and move to the
right ('clockwise')? This was not inevitable. Why not a 24-hour face and
have the clock hands move to the left? If you think this is silly, go to
Florence cathedral and observe its clock move 'counter-clockwise'
through 24 hours. The cathedral and clock date from 1442. At that time
the convention was open. Shortly after, clockmakers standardized on
'our' 12-hour clockwise convention. Yet if 51 percent of clocks had ever
been like the one in Florence, we would now be reading a 24-hour clock
backwards and the clock in the first line of George Orwell's *1984* could
not have shocked readers by chiming 13.

The early bird captures the convention, and hence the competitive
advantage. So get to market quickly, establish the standard, and grab competitive advantage while the field is still fallow.

Business is fractal

There is too much uncertainty and uniqueness in business—in the language of chaos, business is too fractal and too sensitive to initial conditions—to allow for 'paint by numbers' strategies, to slavishly follow well-worn rules of thumb.

This does not mean that we should give up and make most decisions by tossing a die. A large part of the difference between successful and unsuccessful executives and entrepreneurs is the ability to recognize fractal patterns and to make decisions accordingly.

If you spend your lifetime looking at clouds or coastlines, you will make a better guess than most people at whether it is going to rain or which way the coast will run beyond the point you can see. If you have many years' experience and track record at making good decisions in a particular industry and market, the odds are that you will continue to make good decisions by recognizing recurrent patterns—as long as you stay in the same industry and market.

One of the sad things about modern corporate life is how often operating managers, who know their markets well, are over-ruled simply because they are unable to explain or rationalize their instincts to their corporate bosses. A more sensible setup would not require explanations; it would just judge by results.

Recognizing that business is fractal gives you many very important warning and encouraging signals:

◆ Experience and intuition will usually win over analysis, because the analysis can never be precise enough or conclusive. Analysis is of course useful as a supplement to intuition, and, as we have seen earlier, the intuition/analysis dichotomy is in some ways a false one; good intuition is the crystallization of previous analyses, and good analysis is often the exploration of intuitive hypotheses by collecting data. Yet so great is the importance of recognizing fractal patterns and what previously went along with them by way of results—realizing that the position now is more similar to previous situation A than it is to previous situation B, and that situation A led to disaster, for example—

that someone who is a great manager and analyst may well lose out to an apparently lazy but cunning old hand.

◆ At the very least, recognition that business is fractal, and that slight differences in inputs can lead to totally different outcomes, should teach us to be humble when we enter 'adjacent' but new markets. These may look the same as the old markets, and yet be subtly different—with shocking results! If you doubt this, look at the terrible track record of most retailers the first time they venture beyond their home country. Part of the problem is that head office tends to export its own proven formula—proven, that is, in a different market—and brushes aside the objections or alternative models proffered by local managers. The local people can't produce analyses to validate their hunches; head office, apparently reasonably, insists: 'Do it our way until you can demonstrate the logic of yours.' Then the foreign approach fails. No one can say why. The missing explanation is simply that business is fractal.

◆ Whenever you face an important decision, try to find the nearest equivalent situation in your experience or that of colleagues or friends. Do not leap to conclusions. Engage in sober debate about the possibilities. Unless you are very certain, draw up a 'top three' list of possibly similar situations, and what happened next.

◆ Once you have headed down a particular path, be aware that you may have made the wrong decision. Look for the early signs that you are on or off the expected path, just as you would when following a map. If the early signs are not what you expect, you have probably chosen the wrong 'fractal comparison.'

Because there are so many different types of business, where different rules apply, specialized businesses always have an advantage over generalized ones. A specialized unit, company, or market will generally win over an undifferentiated one. Where possible, therefore, form specialized teams, new business units, new divisions, and, above all, new companies. Business is generally reluctant to follow these steps, and increasingly reluctant as the order I have given ascends. A specialized team, maybe; splitting the company into two, no way. Yet if business is fractal, and part of the skill in business is recognizing fractal patterns, a specialized company will be much better placed to recognize the right patterns and make the right responses. A specialized team will go some way toward this benefit, but be constrained by the degree to which it shares resources or decision making with other parts of the organization that deal with different sorts of patterns.

Complexity and emergence

Most of the radical scientific innovators that gave us chaos have now moved on to the more topical study of **Complexity**. Complexity is the study of complex systems that manage to produce their own brand of order. Sometimes this hints at deep simplicity behind the complexity, especially when the same simple yet baffling patterns emerge in completely different types of complex systems; an economic slump, it transpires, is very much like a hurricane, with similar feedback causes and effects; a developing city is very like a growing embryo.

Complexity builds on the insights from chaos, but adds three new themes. First, complexity focuses on complicated feedback systems and shows that these usually have surprising results. Second, as Philip Anderson—widely regarded as the founder of complexity—urged, complexity is about **Emergence**, how groups or 'wholes' behave quite differently from the aggregation of their individual characteristics. In combining individual units—individual customers, water molecules, body cells, business units, single birds, whatever—into groups—into markets, steam, a butterfly wing, a company, a flock—we may emerge with something completely unexpected and different from what we started with. Third, what complexity is really interested in is **Self-organizing systems**: systems that start in a similar or random state but somehow organize themselves, quite spontaneously, into a large-scale pattern.

Self-organizing systems

Spontaneous self-organization is a fascinating thing, especially when the individual components are so numerous and apparently unrelated to each other. Adam Smith (although he didn't know it) was an early exponent of complexity, talking about the 'invisible hand' that seemed to direct the self-interested intentions of millions of producers to satisfy the self-interest of millions of consumers.[6] Think of the billions of interconnected neurons in your brain, producing a result that surely an individual neuron could not envisage, and yet organizing effectively to ensure that you can understand my words. Think of the way that a city that starts out by being racially integrated soon divides itself into racial (or social, or lifestyle) groups; self-organization, incidentally, is not always a benign force. Think of the way that a stock-market crash, or a hurricane, or an earthquake, or a meteorite organizes itself from its constituent parts. Think of how atoms combine into molecules, by forming chemical bonds with each other; the molecules are quite different from, and more complex than, the atoms whence they came.

Or think of the internet. Nobody planned its evolution from being a research tool of government and universities into what it is today: a global network giving information and power to consumers that will trigger the biggest and fastest change in industry and corporate structures ever seen. The internet has a life of its own and decides what it will become as it goes along.

Complex systems are adaptive

Self-organizing systems that are complex are also adaptive. They adapt to their surroundings and try to turn what is happening to their advantage. The brain develops and learns. Species evolve. Cities respond to new inputs. Markets become larger and more specialized, and adapt to pressure from important distant markets.

Complexity is linked to evolution by natural selection. For example, a termite colony adjusts the numbers of its different castes by cascades of chemicals activated in the termite larvae. If there are too few 'soldier termites,' the smell given off by them in a colony falls below a certain level, and then the 'larvae nursery' automatically produces more soldier termites, which differ physically from other termites. John Tyler Bonner has shown how this type of self-adjusting complexity evolves by natural selection.[7] If this happens in nature, is it fanciful to see the emergence of complex systems such as cities and economies as part of the same process of evolution by natural selection?

Complexity theorists such as John Holland explain that complex adaptive systems typically have many niches, each with a specialized role and place. And new niches are always unfolding: niches for new predators, for new prey, for new symbiotic partners, for new parasites. As new niches open up, the system changes. It can never be in equilibrium.

The edge of chaos

Complex systems are perched on the *Edge of chaos*, a curious state between order and disorder, between the status quo and radical innovation, between stability and transformation. Note that complex systems have a combination of order and randomness; they always operate within boundaries, within a structure of order. You can't have self-organization without also having boundaries. When a complex system moves over the edge of chaos, then it crosses a boundary and becomes something different. John Horgen has claimed that 'everything interesting happens at the edge of chaos.'[8]

Biologists use the 'edge effect' to describe the tendency for a greater variety and density of organisms to cluster in the boundaries between communities. In complexity theory, the 'edge of chaos' describes complex systems, because they possess both elements of order and elements of fluidity. A crystal is not a complex system, because it possesses perfect internal order and there is nothing left to change. At the other extreme, a boiling liquid is a chaotic rather than complex system; there is very little order. By way of contrast, a complex system such as amoebae, the stock exchange or an economy has both order and enough fluidity to change. In the words of biologist E O Wilson:

> *The system that will evolve most rapidly must fall between, and more precisely on, the edge of chaos—possessing order, but with the parts connected loosely enough to be easily altered.*

Zipf's rank/size rule

The brilliant economist Paul Krugman has shown that cities behave in many different ways like complex, self-organizing, adaptive systems. Many of his arguments are highly technical, but one that is easily understood concerns the size of American cities.[9] It turns out that they obey **Zipf's rank/size rule**, from Harvard professor of philology George Zipf, which says that the population of a city in any country is inversely proportional to its ranking. If the rule works precisely, then the second largest city would have half the population of the largest; and the third largest would have one third the people of the top city, and so on.

Clearly, we should not expect a perfect fit; this never happens with data and laws. And you might protest that Los Angeles has well over half the population of New York. But as you go down the rankings, the fit becomes amazing. City number 10 in the US is Houston, with 3.85 million people. City number 100 is Spokane in Washington State; this has 370,000 people, fractionally under one tenth the size of Houston. Krugman tells us:

> *If you regress the log of rank on the log of population, you get a coefficient of −1.003, with a standard error of only 0.01—a slope close to 1 and very tightly fitted. We are unused to seeing regularities this exact in economics— it is so exact that I find it spooky.*

Even more eerie is that this power law has worked well for at least a century. The same shape of line emerges if we analyze relative city sizes in 1940 or 1890.[10]

Simon's theory of clumps and lumps

Why do cities organize themselves like this? We don't know. But one idea that Herbert Simon came up with nearly a half-century ago, which may well be right, is that whatever size cities start out, they will then attract to themselves a similar proportion of any subsequent increase in population. This is the **Theory of clumps and lumps**. Each existing city is a 'clump' of whatever size. When population increases, it does so in 'lumps,' not just through the excess of new births over deaths, but through lumps of new arrivals (by immigration in the case of the US; in many other countries by people drifting from the countryside). But each new lump will tend to attach itself to a clump in proportion to the existing size of the lump. Simon explains this by the availability of employment, and by the observed fact that most entrepreneurs (who provide the employment) stay near where they started.

Gutenberg-Richter law

Whether or not you believe the clumps and lumps theory with regard to cities, the weird thing is that Zipf's power law does work, not just for cities, but for things you might think totally dissimilar. It works in the same way for earthquakes, meteorites, and species. Zipf's law applied to earthquakes gives us the **Gutenberg-Richter Law**, which says that the frequency of earthquakes is inversely proportional to their size. Similarly, the frequency with which a meteorite hits Earth is—happily!—inversely proportional to its size. Or if we plot the number of animal species that exceed a particular size, we uncover the same relationship.

Cities, the economy, earthquakes, meteorites, and quite likely evolution too, are self-organizing systems that behave in clear and similar patterns, and that produce order from instability. The whole mysteriously assembles; we are back to Adam Smith's 'hidden hand,' although this paw stretches around far more than the economy. The constituent parts surely cannot 'know' what they are doing. Or can they? In biology, how do we explain how cells arrive in their allotted places? Krugman comments:

> An individual fruit fly cell does not think to itself, 'I am part of a wing', yet cells collectively seem in effect to decide to become different parts of the organism. Experiments suggest that cells indeed behave as if they knew their own polar co-ordinates.

In the business world, think of how teams sometimes magically gel together, and define, without the need for words, each individual's role.

Or, more darkly, how easily a crowd can turn into a frenzied mob, acting in perfect accord to destroy something hated. The whole is more fundamental and purposeful than the parts. Self-organization, and the emergence and adaptation of complex systems, are deeply rooted in the universe, and we had better notice, respect, and take account of them.

Parkinson's laws

One man who did take note of organizations' self-organizing characteristics—although the concept of self-organization had not yet been invented—was C Northcote Parkinson (1909–93). In 1958 he published *Parkinson's Law*, a commentary on organisations that was both serious and satirical. The eponymous law was that 'work expands to fill the time available.'

Parkinson's thesis was that bosses increase the size of their departments because there is more work to do and because they like to have large empires, not because they need to in rational economic terms. 'An official wants to multiply subordinates, not rivals,' he comments; noting also that 'officials make work for each other.' The fact that there is work to do justifies and masks the real objective, which has nothing to do with economic logic.

This law works in the professions also. As has been remarked, 'Town with one lawyer: poor lawyer. Town with two lawyers: rich lawyers.'

Parkinson himself was an official in the British Navy during the Second World War. He pointed out in his book that whereas the number of officers and men in the Navy itself fell by 31 percent between 1914 and 1928, and the number of ships fell even more sharply, by 61 percent, the Admiralty administrators yet contrived to increase their ranks by 78 percent! In complexity terms, the Navy administration was self-organizing, fulfilling objectives that were quite independent of the original intention.

Parkinson later followed up with a second law, applying both to individuals and to corporations: 'expenditure rises to meet incomes.' This idea is closely related to a more academically respectable concept invented about the same time that is also an illustration of organizations' self-organizing tendencies: organizational slack.

Cyert and March's theory of organizational slack

Organizational slack was invented by two academics from Carnegie-Mellon University in Pittsburgh, Richard Cyert and James March, in their 1963 book *A Behavioral Theory of the Firm*. The theory states that firms are not profit maximizing, but rather coalitions of interests, and that firms deliberately built up excess resources during times of success, so that they have fat that they can shed to survive in hard times. The fat is the 'organizational slack.'

Cyert and March applied no normative judgment to organizational slack; as the title of their book suggests, they saw it in behavioral terms and were explaining what happens and why. They would have been very happy with the concept of self-organization.

As businesspeople, however, while observing the tendencies that Cyert and March describe, we may not be so tolerant and disinterested. Fat begets fat; once one department or division is allowed to be overstaffed, others follow fast. Fat obstructs flexibility and speed of response. Fat makes it even more difficult than it always is to be customer centered. So the existence of organizational slack makes it more likely that bad times will come. And the other way round: cutting slack and putting it to work for customers makes it more likely that *competitors will never be able to catch up*, decreasing the chances of severe profit pressure.

The opposite of organizational slack is discretionary investment that builds future profits by improving what is offered to customers. It is sometimes difficult to differentiate between these two opposites: is a large R&D department an investment for the future, or organizational slack? It's probably both. Top management will always say that it is an investment; yet, if that is so, why is the first response when profits fall usually to cut such investments as though they were just organizational slack? The only way to be sure that something is an investment and not slack is to outsource it, so that there is no internal vested interest to fall foul of Parkinson's first two laws and Cyert and March's organizational slack.

Complexity and business

Complexity theory is about complicated feedback systems, about how groups or whole entities 'emerge' from quite different parts, about how complex systems poise themselves on the 'edge of chaos,' and about how they organize themselves spontaneously and deliberately into large-scale patterns. So what?

Be aware that small changes can transform the whole competitive system

Both quantum and complexity theories tell us that reality is in some ways indivisible; in David Bohm's words, an 'undivided wholeness.' The character of the whole arises through the relationships between parts, and when the parts change, then the whole may radically alter its character. If some specialty stores are opened in a shopping mall, this may modestly boost the business of existing stores, by bringing in new customers. But if another mall opens up 15 miles down the road, the stores may suddenly become unprofitable.

Business is intrinsically exposed to such sudden swings from apparently minor influences. How useful is this knowledge? I'm not sure. Of course, we should try to keep an eye open for challenges from unsuspected quarters; but few of us are blessed with panoramic vision. The main lesson may be that when things go wrong, we shouldn't assume that we know why.

One common mistake is to assume that something in the market has changed. Another mistake is to invent a large cause when a small one will do.

Suppose that sales of Filofaxes slump. Why? One explanation is that yuppies are poorer than they used to be, or are dying out, and are therefore not buying Filofaxes. If true, this is a major market change, and there's not much that Filofax can do about it. But what if something much less momentous is happening, like the entry of a new competitor selling cheaper personal organizers? This is a smaller change, and something *can* be done about it. Yet in 1990, Filofax almost went bust because it believed that its problem was a major market shift rather than a minor change in competition.

At the level of the global economy, small events can also have weird and unpredictable consequences. It is not likely, for example, that the Russian economic crisis of late 1998 and early 1999, which nearly derailed the world economy, would have happened if President Clinton had not been totally preoccupied with the little local matter of his possible impeachment.

Complex systems are inherently unpredictable. The global economy is probably the most complex system on Earth, possibly in the whole universe. We should not be surprised that the global economy is so unpredictable or subject to unintended consequences. Yet there are still detectable patterns well known to stockbrokers and others: 'when America sneezes, the rest of the world catches pneumonia.'

Look for and practice emergence

Complex systems come together from the bottom up. They emerge. They evolve. They cohere. They come together, from many constituent parts. They seem to have no problem doing this. Structure comes from no structure, or from lesser structures. The universe manages somehow to bring forth ever more complex structures from simple causes: bacteria, plants, animals, stars, galaxies.

The most important things in the world—like the mind, consciousness, markets, economies, and society—are emergent phenomena. They are not planned. They happen. Once again, we see the 'invisible hand' working overtime.

Recently, the more intelligent writers on business strategy have realized that it, too, should emerge rather than be planned or dictated. Ask yourself this: How did the most successful corporations from, say, 1750 to 1960 arrive at their successful strategies, before we knew how to plan them? Or ask another question: Who knows better what a firm should do, its bosses or the market? The leaders or the troops?

If we reflect on emergence, we realize that experiments of great strategic value often occur as we go along in the normal course of business.

We open a small restaurant and it is amazingly successful. A dull and low-status business unit invents a minor new product that suddenly is all the rage. We botch the technology and stumble over the Post-it® note. Things come together. Success emerges. But the best laid plans of mice and men...

The lesson is not to do nothing and hope for the best. The lesson is to *spot* emergence and then give it one almighty push on its way. It is better to observe the market than to plan it. It is better to see what is emerging from the lower reaches of the firm than to dictate from on high.

Think carefully about the role of self-organization in your organization

Almost the only class of complex system in the universe that is not purely self-organizing is the modern business corporation and other hierarchical organizations modeled on it.

A laser organizes itself: photons (light particles) spontaneously organize themselves into a beam. A hurricane organizes itself. A living cell manages to self-organize. Cities organize themselves. So do economies—when left alone. The Soviet Union is said to have had eight million managers in its economic planning bureaucracy, which proved to be eight million too many.

Why do so many businesspeople preach *laissez-faire* for the economy, but never for the structure of the firms within it?

I'm not advocating that you automatically leave your organizations to organize themselves. The cost of self-adjustment could be much higher than that of intervention. There are powerful selfish and socialistic tendencies built into human behavior and into large organizations. Just because organizations *can* organize themselves does not mean that they *will* organize themselves the way you want them to, or that you have no right to intervene. Resources will probably be wasted by self-organization.

But you should recognize the tendency towards self-organization. Sometimes this should make you stand aside, and allow a team to work out how to do what you have all agreed to do. Sometimes it should make you extremely vigilant, aware that a system that is not controlled or watched will develop its own agenda. Sometimes it should make you avoid complex adaptive systems altogether: if you go with a simpler system, it is more likely to do what you want. If you double the size, or increase the complexity, of your organization, don't be surprised if it does unwelcome new things. That's self-organization!

View your firm as a living organism, as a complex adaptive system

If it has a life of its own, the organization is more than the sum of its parts. It is more than a set of economic transactions. It is more than the people who work in it. It is even more than the set of relationships that it builds. The organization is its own thing; it belongs to its own species. It can breed and it can die. It is a highly ambiguous entity. It can be bought and sold, just as if it were property, a pet or a slave, yet each time it is bought it becomes something new—subtly or not-so-subtly different, yet always recognizably similar to its previous incarnation. It's all a great mystery: difficult to understand, and difficult to describe.

It shouldn't be so hard. As Peter Senge says:

> Is it that we think life starts and ends with us [humans]? Surely, simpler organisms are alive. Why then can't we regard more complex organisms, like families or societies or companies, as being alive as well? Is the tide pool, a teeming community of life, any less alive than the anemones, mussels or hermit crabs that populate it?[11]

What difference does it make if we view the firm as an organism rather than a machine? Here are eight benefits:

- ◆ *It takes away the illusion of control.* A living thing, even a pet or a slave, is more difficult to control than a machine. An organism is unpredictable and headstrong. It has a mind of its own.
- ◆ *It stresses the role of growth and innovation.* Machines don't grow. Organisms can't do anything else (or they die).

◆ *It reminds us that organizations, or parts of them, can be self-starting.* A machine needs to be started, switched on and switched off. Machines suffer from entropy: they run down unless they are regularly maintained. Organisms can start themselves and renew themselves: they grow new cells and regulate their own metabolisms. We could debate, in the case of any particular organization, whether it is more like an entropy-bound machine or a self-renewing organism; clearly, there is a continuum between these two extremes.

◆ *Organisms are part of systems.* An organism is a complex whole composed of many subsystems, and part of many 'super-systems' above it. The corporation is going to be affected by a change in its subsystems—for example by the recruitment or retirement of individuals—and by super-systems—its market and competitive environment. Machines are not affected like this.

◆ *Organisms can build networks and relationships.* Machines can't. Humans, and a few other organisms, can. Admittedly, the parallel is not perfect. It is not so much the organization as such building the networks and relationships, but rather the humans inside the organization building these relationships. There is a danger, therefore, of slipping into 'anthropomorphic' talk about organizations, of regarding them as people. It may be better to regard an organization as a person rather than as a machine, but it is definitely not correct to think that the organization basically comprises its people and is an extension of them. Here we need a good dose of quantum indeterminacy. The people are not owned by the organization or compelled to remain part of it. Even when they are in it, the people have a life outside it. The networks and relationships that are built are not just with other humans, but with other organizations and with society as a whole.

◆ *Organisms have their own purpose.* Machines have the purpose prescribed by their builders or owners. Organizations have purposes that evolve as a result of their founders' characteristics and what happens along the way. Does Microsoft have a purpose aside from making money for its owners? Or the Disney Corporation? Or McDonald's? Of course they do. You couldn't imagine any of these organizations swapping purposes, even if the owners told them to.

◆ *Organisms learn.* Only living things can learn. Clearly, organizations can learn: Greenpeace can learn, a rock band can learn, a baseball team can learn, Microsoft can learn (about the importance of the internet, for example, even when the company's founders were sceptical), even the US Republican and Democratic political parties can learn! It is worth asking, however, whether organizations can know more than their individual members know in total. Knowledge may exist as a function of working together, but knowledge can surely only reside within the human players, not in the organization independent of its members.

As Peter Senge notes,[12] it is not just organizations that learn, but also the global business community. Here is yet another society or organism. Technologies and ways of doing business get copied and extended. Self-service, the multidivisional corporation, the multi-national corporation, hostile takeovers, leveraged buyouts and buyins, and spin-offs are invented in America and exported to most other economies. Total quality management is elaborated in Japan and then reimported into the United States, and, within a few years, it's ubiqui-tous. (As we saw in Chapter 1, organisms mutate and species learn and improvements are diffused rapidly, because there are only those that learn and those that die. In this respect, machines have an easier time.)

◆ Finally, *organisms can have their own character and uniqueness.* It is possible for a machine to be unique, but it's highly unusual. A machine that has its own characteristics rather than those intended by its designers is proba-bly not a very good machine. On the other hand, organisms are unique, and sophisticated organisms do exhibit their own character. Humans, and possibly other organisms, have emotions. Organizations have their own cultures that are the product of history and accident as well as human design. A firm is *sui generis*, of its own species. One important character-istic of the species is that each member of it develops its own unique way of doing things, and emotion is an important part of this.

Use the power of 'landscapes'

One of the most useful concepts to emerge from complexity science is the metaphor of '**Landscapes**.'[13] Michael Lissack and Johan Roos point out that we are hardwired to recognize patterns in space. They provide a thrilling account of developments in the PC business in terms of landscapes (see also their depiction of this on p210):

> *As technology shifted, the hill which was best to climb shifted as well. If you picture time as a landscape of clay made by a child, then the history of major events in the PC world is similar to what would happen if an angry adult came along, picked up the model and gave it a good, hard twist. Up pops a hill where one wasn't before, and the old hills seem to fold in on themselves. IBM once owned 25 per cent of Intel, all of Windows, and had the opportunity to buy both Microsoft and Apple...*
>
> *Picture a set of mountains, each a distinct subset of the competitive world. IBM on one, Xerox on another, AT&T on a third. They are connected only by difficult mountain passes, and the only transport is donkeys. Suddenly, two changes occur. The Swiss tunnel through the mountains and the automobile is invented ... Just yesterday trekking from IBMland to Xeroxland was the journey of a lifetime, now it is summoned up by a button on your desk and can be ended in an instant. Where is the landscape now?*
>
> *The graphical interface introduced by Apple (that little outpost) would now seize ... the hill. IBM wanted a graphical interface for itself ... In the process, it created Bill Gates ...*
>
> *Now the landscapes were merging as well as shifting. IBM's mountain was being eroded from underneath...there was a volcanic eruption ... Intel and Microsoft, two lowly suppliers, were transforming the landscape ... With Windows 3.0, Mount St Helens erupted. IBM's hilltop was no more. The hill itself was still there, but it was lower by a few thousand feet and had a large crater in its midst.*

Create your own landscape metaphors to describe what is happening, and what might happen, in your own terrain.

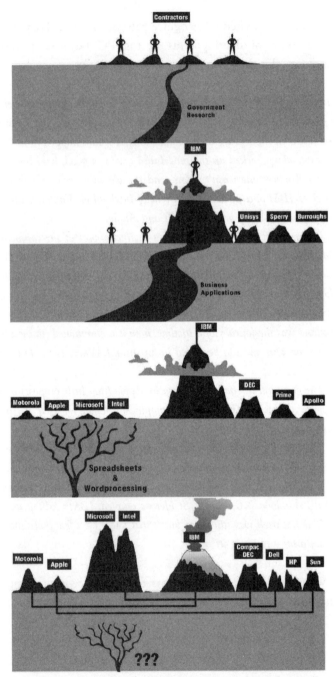

Source: Michael Lissack and Johan Roos (1999) *The Next Common Sense: Mastering Corporate Complexity Through Coherence*, Nicholas Brealey Publishing, London. Reprinted with permission.

Summary

Chaos tells us that events unfold with 'sensitive dependence on initial conditions.' We may not know what the initial conditions were, and we may therefore attribute important results to luck. But there is a pattern in how events unfold. For executives, the two most important corollaries of chaos are:

◆ In any embryonic market that has the potential to become large, grab the first-mover advantage.
◆ Business is fractal. Each type of business has its own patterns, endlessly varied, endlessly similar, never precisely repeated. Success comes from recognizing the patterns. This requires specialization, experience, and an instinct for the patterns themselves.

Complexity tells us that small changes can completely reconfigure large and complex systems. Big changes can have small causes. Structure is not planned; it emerges on its own. Given half a chance, complex systems organize themselves. In the case of organizations, this will lead to unexpected consequences, which may not always be welcome to owners and leaders.

Action implications

◆ *Exploit the fractal nature of business*. Gain experience and skill in spotting the recurrent patterns that are peculiar to your own business space. Remember that it is experience at pattern spotting, not the time that you spend in a market, that is valuable. Don't expect the patterns to be the same as in other markets, even if they are very adjacent and appear very similar.
◆ *Grab the first-mover advantage*.
◆ *Be flexible and have strategies, not just one strategy*. Realize the role of chance and apparently trivial events. Build flexibility into your plans and actions. Be willing and able to change tack halfway. Allocate some of your resources to 'long-odds' bets: experiments that will probably come

to nothing but could contain the seeds of major success if market conditions change radically.

◆ *However unfavorable the circumstances, look for the things that you can control.* When things go wrong, don't invent the excuse of a large and uncontrollable market change. Look for all the small things that could have gone wrong and that you could influence.

◆ *Practice emergence.* Identify emergent trends and unexpected successes, and go with their flow.

◆ *Don't overorganize, but don't let self-organization scupper your plans.* Be vigilant for evidence of corporate self-indulgence.

◆ *Expect non-linearity.* Don't expect simple cause-and-effect relationships to predominate. Find hidden, recurrent patterns within apparently senseless data. Immerse yourself in the data and the action and leave your brain to sort out useful patterns.

◆ *Use landscape metaphors to describe your industry and to visualize alternative futures.*

Notes

1   Quoted in James Gleick (1987) *Chaos*, Little, Brown, New York.

2   Henri Poincaré (1908) *Science et Méthode*, Ernest Flammarion, Paris; quoted in David Ruelle (1991) *Chance and Chaos*, Princeton University Press, Princeton, also published (1993) by Penguin, London.

3   Although it quickly became famous, Lorenz's paper was initially not published, except as a press release of the conference of the American Association for the Advancement of Science, to which the paper was presented on December 29, 1972 in Washington, DC. The butterfly paper was first published, along with other lectures, in Edward Lorenz (1993) *The Essence of Chaos*, University of Washington Press, Seattle.

4   Erik Durschmied (1999) *The Hinge Factor: How Chance And Stupidity Have Changed History*, Hodder & Stoughton, London.

5   Eric D. Beinhocker (1999) 'On the origin of strategies,' *McKinsey Quarterly*, 4, pp 47–57.

6   Adam Smith actually identified several ways in which the economy automatically adjusted itself in accordance with what we would today call 'feed-

back mechanisms.' He showed that high prices stimulated production of those goods and low prices discouraged production, thereby matching supplies more closely to demand. He also showed how the prices of wages and capital triggered desirable adjustments.

7 John Tyler Bonner (1988) *The Evolution of Complexity by Means of Natural Selection*, Princeton University Press, Princeton.

8 John Horgen (1995) 'From complexity to perplexity,' *Scientific American*, June.

9 See the terrific little book by Paul Krugman (1996) *The Self-Organizing Economy*, Blackwell Publishers, Cambridge, MA.

10 Krugman notes that Zipf's law does not work so well in countries with a single pre-eminent 'primate city' that combines the 'normal' economic role with that of the political center; places like London or Paris. To the expected 'economic' population, we have to add the employment provided by the bureaucracy and all those who cluster around power. Zipf's law then works with hypothetically adjusted populations. It also works in most countries where there is no primate city.

11 Foreword to Arie de Geus (1997) *The Living Company: Growth, Learning and Longevity in Business*, Nicholas Brealey, London.

12 Peter M. Senge (1994) *The Fifth Discipline: the Art and Practice of the Learning Organization*, Doubleday, New York, p. 4.

13 See the excellent account in Michael Lissack and Johan Roos (1999) *The Next Common Sense*, Nicholas Brealey, London.

10

On the 80/20 Principle

For a very long time, the Pareto law [the 80/20 Principle] has lumbered around the economic scene like an erratic block on the landscape: an empirical law which works and which nobody can explain.

Josef Steindl, economist

A microchip's physical content isn't very valuable. Silica is the cheapest and most abundant raw material on the planet—sand. But a microchip—its shape, its design, its unseen artistry—is extremely valuable. Yet it comes from a source that seems almost unlimited—the knowledge and inspiration that we draw from the human mind and spirit. This is the most valuable resource and the most abundant.

Tachi Kiuchi, chairman, Mitsubishi Electric America

More with less

The history of civilization is the history of achieving more with less: the happiest non-linear relationship. The development of agriculture some 7,000 years ago moved mankind beyond Stone Age hunter-gathering. The progress of science since the seventeenth century enabled fantastic and sustained leaps in agricultural and industrial productivity that have enabled the earth to support quite unprecedented numbers of people, hundreds of millions of them at or beyond living standards historically reserved for a tiny élite. Science and technology have enabled us to do things—produce food and shelter, conquer disease, travel, build monuments, communicate, create art, enjoy ourselves—to progressively higher standards using only tiny fractions of the natural resources and time that used to be necessary.

Reflect that two or three centuries ago 98 percent of the labor force was employed on the land, producing food, and now 2–3 percent of the workforce produces far more food for far more people with far less effort. Then think of how much is produced by computers and the internet, and how few natural resources go into the process. A century ago there were no computers. And as Diane Coyle comments,[1] a single birthday card weighing less than a gram with a microchip that plays 'happy birthday' when you open it contains more computer power than existed on the whole planet 50 years ago.

More for less, the hallmark of progress and wealth creation throughout the ages. More for less is always possible. It's also, happily, quite inevitable. Sooner or later, in anything, we will get more for less. And however much more you get for however much less today, you can be absolutely confident that tomorrow you will get even more than today's more for even less than today's less.

Furthermore, it is all down to science and technology, to intelligent use of knowledge.

A myriad of specific applications and improvements incarnate science's forward march. But it is also worth asking whether there are general laws that underpin the process, that can help us extract more from less in *any* application or task.

It so happens that there are, and this chapter explores them.

Language, the movies, and the world wide web

What do language, movies, and the world wide web all have in common? Many things, perhaps, but one key commonality is that they are all highly pronounced examples of an extremely useful power law, one that can always point the direction to achieving more with less.

In 1999, two Xerox corporation researchers[2] found that a tiny proportion of sites on the world wide web command most of the traffic: 119 sites—fewer than one tenth of 1 percent—received 32 percent of all visits (the top site was Yahoo!). The top 5 percent of sites in the sample, comprising about 6000 sites, received 75 percent of visits.

The same pattern of a few important winners and a mass of unimportant losers is evident in the movies. In 1997, two economists studied the revenues and lifespans of 300 movies released over an 18-month period. They found that four movies earned 80 percent of the box office, while the other 296 movies had to make do with a miserable share of the remaining 20 percent.[3] In other words, 1.3 percent of the total number of movies accounted for 80 percent of the revenue: an even more extreme example than the world wide web, but the same general skewed distribution.

The third example is everyday language. Sir Isaac Pitman invented shorthand after finding that just 700 words made up a staggering 70 percent of conversation. Including derivatives of these words, Pitman found, the proportion went up to 80 percent. The *New Oxford Shorter English Dictionary* lists more than half a million words. What this means, then, is that fewer than 1 percent of words make up 80 percent of word usage. This pattern is very similar to that of the movies.

The 80/20 principle

What the web, movies and speech have in common is a profound imbalance in how the spoils are divided. The first person to spot the prevalence of such patterns was the Italian economist Vilfredo Pareto in 1897, when he looked at the distribution of wealth and income across the working population.[4] Pareto found a small minority earned a substantial majority of total incomes (or enjoyed a predominant share of wealth), and what really excited him was that the distribution followed almost exactly the same pattern whatever the time period and also whichever country he examined. In the last half century, Pareto's law has come to be generally known as the **80/20 Principle** (or 80/20 rule) based around the rough observation that the top 20 percent of any distribution usually accounts for about 80 percent of its power or impact.

In business, many studies have shown that the most popular 20 percent of products account for approximately 80 percent of sales; and that the 20 percent of the largest customers also account for about 80 percent of sales; and that roughly 20 percent of sales account for 80 percent of profits. Likewise, it is a safe bet that about 80 percent of crime will be accounted for by only 20 percent of criminals, that 80 percent of accidents will be due to 20 percent of drivers, that 80 percent of wear and tear on your carpets will occur in only 20 percent of their area, and that 20 percent of your clothes get worn 80 percent of the time.

'80/20' is not a magic formula. The actual pattern is very unlikely to be precisely 80/20. Sometimes the relationship between results and causes is closer to 70/30 than to 80/20. Sometimes, as in the three examples above, the pattern is even more extreme than 80/20. For the world wide web, there is a 75/5 relationship: 5 percent of sites attracted 75 percent of visits; and about 7 percent of sites receive 80 percent, so in this case it is 80/7 rather than 80/20. For movies it is 80/1 (to the nearest round number): 1 percent of movies make 80 percent of the box office gross. The use of words also displays an 80/1 relationship: fewer than 1 percent of words is used 80 percent of the time. The point is not whether a particular example is 80/1, 80/7, 80/20 or 80/30.

The point is that *the relationship between causes and effects is very rarely 50/50 or anywhere near it.* In this sense, the universe is not very democratic; neither is the world wide web, despite hopes that it would allow a large number of contestants to compete on a level playing field. There is nearly always a 'sheep and goats' phenomenon: some parts of the picture are hugely important, and most of the rest is insignificant background. The power of the 80/20 principle lies in the fact that it is not fully intuitive. Although we do expect some things to be more important than others, we don't expect the differences between the important things and the less important things to be anywhere near as great as they usually are.

The universe is predictably unbalanced—roughly along 80/20 lines. To a much greater extent than we expect, few things really matter. *Truly effective people and organizations batten on to the few powerful forces at work in their worlds, and turn them to their advantage.*

Less is more

'Less is more' was made famous by Ludwig Mies van der Rohe (1886–1969), the 'minimalist' German architect, who meant that an architect should concentrate on the essentials of buildings and not try to hide the materials from which they were constructed. The phrase actually comes from Robert Browning's 1855 poem, *Andrea del Sarto*.

'Less is more' is a useful catchphrase because it reminds us that much of what we do actually has negative value. Many activities, customers, products, and suppliers actually *subtract* value, which helps to explain why their very positive counterparts produce such a high proportion of net value. So here's another even more useful motto for businesspeople, courtesy of Bill Bain, founder of consultants Bain & Co: 'The best way to start making money is to stop losing money.' The best way to become more effective is to stop your negative activities. I will elaborate on this theme under the heading 'Trichotomy law' at the end of this chapter.

Zipf's principle of least effort

In 1949, George Zipf laid out his **Principle of least effort**, which was actually a rediscovery and elaboration of Pareto's law. Zipf's principle says that anything productive—people, goods, time, or skills—naturally tend to organize themselves so as to minimize work. This explained his finding that approximately 20–30 percent of any resource accounts for 70–80 percent of the activity related to that resource.

Juran's rule of the vital few

One of the great heroes of the last century—although with good reason more honored in Japan than in the United States, the land of his adoption—was Joseph Moses Juran. More than anyone else,[5] Juran pioneered the quality revolution that made the second half of the twentieth century a time of unprecedented, and increasingly global, advances in the quality of consumer products from cars to computers. In 1951, Juran published the first edition of his *Quality Control Handbook*, which made what he alternately called the Pareto principle and the **Rule of the vital few** synonymous with the search for dramatically higher product quality.

Juran said that the key quest was always to isolate the 'vital few' causes of anything—in his case, of poor quality—as opposed to the 'trivial many.' Quality losses were not, in general, due to a multiplicity of causes. In each case, a few vital causes could be identified.

Juran made little headway in America in the two years after he published his great work. But his 1953 lectures in Japan caused a sensation. He stayed on to work with several major Japanese corporations, causing them to approach, then to catch up, and finally to exceed the best American quality standards. It was only in the 1970s and 1980s, when Japanese competition menaced Europe and America, that Juran and his movement were taken seriously in the West. He moved back to do for American business what he'd achieved in Japan.

Giant strides in computing using the 80/20 principle

In 1963, IBM realized that about 80 percent of a computer's time was spent on a maximum of 20 percent of the operating code. This insight immediately led the company to rewrite its software to make that most popular 20 percent much more accessible, fast, and user friendly than previously, giving it a significant advantage over competitors, who for a long time continued to treat all applications more or less equally. In the 1990s Microsoft took the 80/20 approach yet further, devoting obsessive energy to simplifying the most popular uses of the PC.

Throughout the industry, most software writers and computing executives became aware, implicitly or explicitly, of the 80/20 principle. Is it a coincidence that, of all high-tech products, the PC is the easiest for technically challenged people like me to master?

Winner take all ('superstar') principle

One illustration of the 80/20 principle is in the massive and increasing gap between the returns of the top earners—whether these are Steven Spielberg, Bill Gates, Rupert Murdoch, Oprah Winfrey, Pete Sampras, Luciano Pavarotti, or the top trial lawyers, writers, and other professionals who are the stars of their own worlds—and those who are just below the top rank. The superstars take an amazing proportion of the total, and their popularity becomes self-reinforcing.

Whenever markets operate freely, they tend to divide the world into a few very fortunate people on the one hand, and everyone else on the other. During the 1980s, an astonishing 64 percent of the total increase in incomes in America went to the top 1 percent: a 64/1 principle! This can be neither healthy for society, nor sustainable, but it does illustrate how free markets, and the universe generally, operate.

I might as well say here that the 80/20 principle, like all power laws, may not be benign. It is simply a force to be reckoned with, which we must turn to our advantage wherever possible. The extent to which we use the 80/20 principle to increase our own effectiveness—by mimicking the way the universe works—may in fact determine the extent to which we are able not only to extend its benign 'natural' effects (for example, in reaching higher and higher levels of productivity and wealth), but also to control and reverse its malign 'natural' effects, which, when allowed to operate unchecked in society, will tend to undermine democracy.

The ubiquity, universality and usefulness of the 80/20 principle

Of all the power laws in this book, the 80/20 principle is one of the most universal. It seems to apply to almost anything. It is built into the fabric of the universe. In one important sense, it is how the universe works and progress occurs. Evolution by natural selection can be viewed as one (tremendously important) subset of the 80/20 principle. (And, if Pareto had written a century before Malthus rather than a century after, it may be that Charles Darwin would have had his insight into natural selection after reading the former rather than the latter. The theory of natural selection actually follows much more directly from the 80/20 principle than it does from Malthus's theory of competition among individuals for food.[6])

Today, we take it for granted that we can compare two related sets of data—such as the distribution of incomes, and the distribution of the people earning them—and observe the disparities. So if we find that 80 percent of total income goes to 20 percent of people, we may not be particularly impressed; we may say, so what? It was Vilfredo Pareto's genius to make comparisons like this for the first time. But what is awesome and spooky is how prevalent the pattern of predictable imbalance is, when applied to almost any two sets of related data.

In evolution, in business, in society, and in life generally, including our personal lives, there are always a few powerful influences, a few things that really matter—and also an enormous amount of background noise, which claims our attention and distracts us, but which is best ignored because it doesn't matter. In paying attention to the background noise, which persuasively masquerades as important, significant and urgent, we limit our effectiveness and squander the energy that should be devoted to observing and co-opting (or avoiding) the powerful forces around us.

It is easy to concede that the 80/20 principle operates across the broad canvas of life; we can hardly deny it when we look in detail at the facts of any particular case. Yet nothing is more difficult, as I have found myself, than to keep remembering that, beneath the hurly-burly of ordinary life—when we are continually assaulted by demands on our attention and time—the 80/20 principle is still operating, and requires a very selective response if we are to be effective. We may know that the 80/20 principle applies, and yet behave as though we didn't.

Virtually all businesses do more than they should, own more than they should, acquire more than they should, and try to exert influence where it is fruitless. Nearly all executives try to manage more than they should, do too many things, and know too much about too much, and too little about the few things that will determine their success or failure. So do the great majority of managers in non-business organizations, and civil servants and politicians. So do almost all of us in our private lives: we spend time, energy and money on things that will only marginally affect our happiness and value to others; we fail to give due weight to the few people, events and objectives that give our lives meaning.

How to use the 80/20 principle in business

There are many helpful *tactical* uses of the 80/20 principle to help your organization or your career—for instance, it can be used in negotiations, including to get yourself a pay rise—and also in your personal life, but since I don't want to repeat what I've said in an earlier book,[7] I'm going to concentrate here on the *strategic* ways to use the principle in business.

The key insight is that your firm almost certainly does too much. The hypothesis is that 20 percent of what it does leads to 80 percent of the benefit. If this is true, it follows that the firm should do more of the 20 percent (or similar activity), but very little of the 80 percent. The firm should do *much less*.

Too abstract? Let me be more concrete. Your firm should do less, but more profitably. Which ever way you cut it, the firm should concentrate on its most productive and profitable elements and activities, and forget or hive off the rest.

So what should it do less of? Try these for starters:

◆ The firm should own less.
◆ The firm should acquire less, and divest more.
◆ The firm should try to participate in fewer stages of the value chain.
◆ The firm should have fewer products.
◆ The firm should have fewer customers.
◆ The firm should have fewer suppliers.
◆ The firm should have fewer employees.

Own less

Managers like to own things. These might go up in value, and it's generally believed that ownership enables us to control what is owned. Alas, very often the reverse is true, both in business and life: our possessions end up controlling us. In business, it is now clear that the obsession with owning things is passé. We don't need to own things to control revenue and profit streams; and there are severe disadvantages to ownership, when compared to the non-ownership options.

By definition, half of the world's total business assets destroy value. They fail to earn the average return on capital. Having them therefore destroys value.

If this sounds too much like a theoretical construct, look at the following examples of value creation and capture without very much ownership.[8] As I write, British pay-TV channel BSkyB has a market value in excess of £10 billion, yet it has a small asset base and broadcasts programs that it does not create or own. Or take Canon and Microsoft. As strategy professor Marcus Alexander writes:

> Canon's dominance of the inner workings of the fax machine, or Microsoft's position as an arbiter of PC standards, effectively capture a disproportionate amount of value through minimal but selective ownership.[9]

Similarly, McDonald's, and a huge number of other businesses, control their suppliers—they tell the suppliers precisely what to do, and often have exclusive arrangements—without the need for any ownership.

This is one meaning of the increasingly prevalent term 'virtual company.'[10] Branded 'manufacturers' of autos and PCs are usually nowadays nothing of the sort; the truth is that they contract out not just manufacture, but often large chunks of design and subassembly. They don't own, yet they do control. Some airlines approach virtuality: they lease their planes; they buy in their engineering, maintenance, catering, ground support services, ticket sales, and in some cases even their pilots and cabin crew. You trust that they control these activities—you trust the brand—yet they are not owned. Whole industries like oil exploration and production are becoming increasingly virtual. Even government is discovering that it can control the delivery of welfare services without having to own them.

There are four advantages to controlling without owning. One is that it takes less cash, so return on capital can go up, and in some cases reach astronomical levels (the extreme is the professional service firm that can own almost literally nothing, and yet control and deliver a highly branded, highly differentiated and highly valued service). The second advantage is that you can focus effort on less activity and become really superb at your narrow speciality; conversely, if you own something, you have to pay some attention to it. Third, if you can control rather than own, you get instant access to the people who are best at each activity. Try comparing this to what you can get in-house. Often, the outside supplier can deliver a better product cheaper than the in-house alternative, and make a fat profit into the bargain. Finally, and perhaps most importantly, controlling without owning leads to flexibility, speed, and the avoidance of unnecessary surpluses paid to internal staff. If you own a division or activity, it creates what Marcus Alexander calls 'ownership inflexibility':

> *Organizations typically evolve around the needs of certain dominant activities or processes. This creates inflexibility in accommodating the different needs of less central processes ... this can be seen in the inflated wage rates historically paid to unskilled or semi-skilled workers [in the oil, chemical and pharmaceutical industries], where the dominant processes involve better paid and hard-to-attract professionals.*

Inflexibility is also apparent when owning assets, such as bank branch networks or production plants based on obsolete technology or high-cost labor, that actually prevent firms from adopting the solutions preferred by customers. Ownership may enable you to control today's processes at the expense of missing out on tomorrow's.

Since knowing how to do something for customers better and cheaper is the only secure basis for companies to make a living, and since knowledge resides ultimately in people, and since slavery has been abolished, I could argue that it is impossible to own the most important components of corporate success in any case. Similarly, many activities that appear to involve ownership need not actually do so. Buildings, computers, communication lines, manufacturing equipment and almost everything else that appears to give substance to companies can be leased or hired.

Own only the 20 percent that contributes 80 percent of effective control. And if control is possible without any ownership, don't own anything.

Acquire less

It is a paradox that profits can be produced without ownership, and yet the easiest way for managers to spend enormous amounts of money is to acquire other companies, which themselves may own very little. Why use capital to acquire firms when the underlying operation can be conducted with very little use of capital?

A partial explanation is that access to a stream of profits is worth paying for. This is only a partial explanation, because it can't explain why acquisition is more prevalent than building businesses from scratch (where much less capital can generate much more cash in the long run), or why the prices paid for acquisitions (as measured by price/earnings ratios) have escalated way beyond sensible calculations of value. Nor can the 'stream of profits' explanation tell us why we should use scarce capital to pay shareholders of an existing business (who themselves have probably provided far less capital than the business is said to be 'worth') when a higher long-term return is likely from organic expansion using little capital.

The paradox can only be explained, I believe, by what I call the 'false market in acquisitions.' Acquisitions are more expensive than is economically justified only because managers prefer to buy than to sell, and because managers have a shorter time horizon than is good for the owners of the businesses (or anyone else). Building companies takes too long for managers to reap the benefit personally.

In the market for companies, there are more buyers than sellers; or rather there would be at economically sensible prices. Inflated prices are necessary to reach equilibrium, where the number of buyers and sellers is equal. This does not mean that all acquisitions are stupid or too expensive; only that the average acquisition is (and all the below-average ones are). Conversely, not all divestments add value; but well over half do.

Mergers and acquisitions (M&A) are still growing (although if you deduct spin-offs, demergers and unbundlings from the total this is not true). But even accepting this growth, the alternative form of combination—corporate alliance—is growing even faster.[11] Corporate alliance is often a better alternative to acquisition.

The global auto market now resembles a tangled mass of spaghetti, a complex network of relationships. Alliances are also the dominant trend in the financial services industry, in computers and in telecommunications. Nor need alliances be confined to one industry. Coca-Cola, McDonald's and Disney have a global alliance that benefits them all, combining Coke's brand and marketing, McDonald's distribution strengths and Disney's branded characters. Virgin uses its brand to enter other businesses, but often without doing anything else, and certainly without acquisition.

It is often easier, and nearly always much cheaper, to get what you want from another company via alliance rather than acquisition.

Participate in fewer stages of the value chain

The value chain is all the activities that lie between the conception of a product or service and its arrival in the hands of the customer. So it involves research and development; product design; component manufacture, assembly, and processing; branding and marketing; selling; physical distribution and delivery; after-sales service, and any other stages that are relevant to your own industry. If, like most businesses, you participate in more than one of these stages, you are unlikely to be equally good, when compared to the best competitor, in both or all stages.

Many very successful businesses focus all or nearly all their energies on one (or two) stages of the value chain. Companies that just do oil exploration, or production, or marketing have tended to be more profitable than the integrated majors. Firms like Filofax that used to undertake substantial elements of product manufacture are more profitable now that they have focused on product design, branding, and marketing. In baby buggies and strollers, companies that just brand and market, *or* that just manufacture, are more profitable than the firms that do it all. Hotel corporations are dividing themselves into units that own and manage property, and units that operate hotels. The profits of the Coca-Cola

Company leapt after it divested its bottling and physical distribution operations.

Concentrate on the 20 percent of activities where you add 80 percent of the value.

Have fewer products

Examine your product profitability. A good hypothesis is that 20 percent of product will generate 80 percent of profits. By definition, the bottom half of products are depressing your average return on capital. It's also likely that the bottom half of products don't meet your required rate of return, and possible that a good chunk of them are actually loss-makers.

If your products conform to the typical pattern, a 20 percent increase in sales, of the most profitable products, would lead to an 80 percent increase in profits. Even a 10 percent sales increase, if concentrated in the most profitable products, will probably lead to a 40–60 percent profit increase. Conversely, a 10–20 percent drop in the least profitable products is likely to lead to an increase in profits, even if no overhead is cut. If overhead can be cut in line with sales, then over a year or so you may be able to cut sales by a third and increase absolute profits, as well as dramatically increasing the return on capital.

Have fewer customers

The same logic applies to customers. Sometimes customers are more (or less) profitable because of the mix of products that they take; sometimes because they pay higher (or lower) prices; but very frequently because they take differential amounts of energy and cost to serve. In most businesses today, the majority of internal cost lies in that elusive beast, 'overhead.' Overhead is very rarely allocated to customers, yet when this is done, even very approximately, it becomes apparent that some customers require a great deal more overhead cost than others, relative to the sales that they take.

Generally, the most profitable customers are those that have been customers for a long time. Gaining new customers is very expensive. If they are not suitable in the first place, or if they are lost quickly, the cost of acquiring these customers may be much greater than the benefit from having them.

It follows that the best customers are generally your existing customers, and that efforts to retain and increase their business with you will have a disproportionate effect on current profits, and, even more importantly, on future profits. Retaining and selling more to the most profitable existing customers has an enormous value that will never be apparent from conventional accounting reports.

Losing your worst customers means that you can provide a better service to your best ones and others like them.

Have fewer suppliers

It is easier to think of customer or product profitability than of supplier profitability, but supplier profitability is just as important and just as differential. Take 10 suppliers. Assume that each charges you $1 million a year for supplies. The 80/20 principle supplies an intriguing hypothesis: 80 percent of the real value is contained in 20 percent of the supplies. If we assume that your return on sales is 10 percent, and that this applies equally to bought-in and in-house activities, it follows that the real value to you of the $10 million of bought-in goods and services is actually $11 million. The hypothesis is therefore that $2 million of the supplies are worth $8.8 million (80 percent of $11 million) and that the remaining $8 million of supplies are worth only the remaining $2.2 million.

Wouldn't it be nice to identify the supplies where each dollar buys you $4.40 worth of value? Wouldn't you tend to buy more of these supplies? Wouldn't you have a strong incentive to sell more of the products and services that are particularly intensive in their use of the profitable supplies?

Wouldn't it also be useful to stop making large losses on the majority of supplies bought in, if the hypothesis turned out to be correct?

Actually, I can't prove that it is. There have been thousands of studies of customer and product profitability that have reproduced the rough 80/20 pattern. But because it is much more difficult to measure supplier profitability, both conceptually and practically, than it is to measure customer or product profitability, I can cite no empirical studies to back up my point about suppliers. Here is a gap in the product line of management consultants that some enterprising firm should plug.

I am confident, however, that some supplies *are* much more profitable than others. There have been studies showing that firms with fewer suppliers are more profitable than comparable firms with more suppliers.[12] One reason for this is that simplicity has high value; but another reason is that firms with fewer suppliers are likely to have deliberately picked the most profitable ones.

Another cause of widely different supplier profitability is the disparity in their bargaining power. Some suppliers create a terrific amount of value but capture only a small share of it, either because the supplier is small relative to the customer, or because the supplier has few possible customers, or simply because the supplier is not maximizing what it could take. At the other extreme, some suppliers almost certainly take more than they give, either because they are skilled in value capture, or because the cost to the purchasing organization is low, or because the purchaser has poor information.

Have fewer employees

This is a touchy subject, but it has long been apparent—both to objective academics and to reflective managers—that in every organization, some individuals add a great deal more value than they extract, whereas for other individuals the reverse is true. Where we can measure individual productivity easily, as with salespeople, the 80/20 principle has been validated: both 80 percent of sales, and 80 percent of the profits from sales, are generated by roughly 20 percent of salespeople.

What is true of individuals in any function or activity is also true between groups. In every organization there is a small corps of individuals, in one particular type of function or activity, who generate the most spectacular profits relative to their cost, and a majority of people in other functions and activities who add little value beyond their cost, or who cost more than they are worth. In the consulting firms where I worked, for example, the real value was added at the top and the bottom of the firm—at the level of the best partners, and at the most junior professional level, where young analysts were cheap, bright and incredibly hard working. The partners sold, and thought; the young analysts did most of the work. The run-of-the-mill consultants pretended to do something useful in between, and were expensive.

In most pharmaceutical companies, it's a few boffins in the labs who add most of the value. In Microsoft, it's probably Bill Gates, a few of his very top executives, and a few of the most creative software nerds. In investment banks, the real profits are generated by the few traders who consistently call their market right, the few analysts who pick excellent investments for the bank's own balance sheet, and the few rainmakers who bring in the mega-deals. In all these organizations, every other cadre and every other individual is seriously overpaid, just because they're there and the organization can afford to overpay them.

Keep the 20 percent (or fewer) of employees who add most of the value, and work out how to export the rest, whether by outsourcing, spin-offs, natural wastage, or less agreeable means. If this is impossible, form a spin-off comprising the most valuable cadre of individuals.

The simple firm

The changes I am recommending have a multiplicative effect, even if you only take small steps in each. Even if each change is modest, the firm that owns less, acquires less, divests more, reduces its scope in terms of value-added activities, trims its product line, and reduces its number of customers, suppliers and employees, will find itself much less complex than when it started. Complexity adds cost; it also makes executives and the whole firm slow, inward looking, and deaf to customers. Simplicity changes the firm's ratio between useful cost and useless cost. Complex organizations spend most of their energy on their own processes. Only simple firms can devote most of their effort to doing useful things for customers.

The firm that does more with less is not necessarily a small firm. Because it does more, the firm may end up being very large indeed, or at least very valuable. The most valuable firm in the world is Microsoft, which is not a small firm. Yet it is a *simple* firm. Simple is beautiful.

Return on management employed (ROME)

Probably the defining, most important business concept of the twentieth century was ROCE, return on capital employed. The defining business concept of the twenty-first century could prove to be ROME,[13] **Return on management employed** (or return on management effort). ROME is increasingly more important than ROCE. The scarce resource is focused, insightful, value-adding executives, not capital. Ideas, brains, knowledge, technical skills, and the sheer ability to get sensible things done— these have more value and rarity than cash.

Whether it knows it or not, the simple firm has worked this out. It is trying to maximize ROME, not ROCE.

More complex firms would do well to classify all their business segments on a two-by-two matrix, looking at ROME and ROCE. Only businesses with high ROCE and high ROME should be retained as part of the core; the rest should be spun off or sold. *Even businesses with high ROCE are not really as profitable as they look if they also have a low ROME.* They are hogging the scarce resource. Financial controllers and CFOs are looking out for high ROCE, and are parsimonious with cash. Who is looking out for ROME? Who is ensuring that it is doled out parsimoniously, only to businesses with really good returns?

Focus on the 20 percent of businesses that have high ROCE and high ROME, the few that really do offer great returns on what is truly scarce. If there is a conflict between ROCE and ROME, give priority to the latter.

Charm

In the 1950s, particle physicists working in quantum electrodynamics (QED) began developing 'electroweak theory.' In the 1970s, Sheldon Lee Glashow (born 1932), son of a Russian immigrant to the US, proved that in addition to the subatomic quarks called 'up,' 'down,' and 'strange,' there was a fourth quark, dubbed '*Charm.*' The theory explains how particles interact with each other. Strong subatomic forces, apparently, require weak forces as a necessary complement.

Sociologists have developed the parallel theory of the power of weak ties.

The power of weak ties

A weak tie is one where there is no direct ownership, financial interest, control, contract, or affiliation; but where there is some connection as a result of knowledge, indirect links ('a friend of a friend'), geography, professional group, or some other accidental or incidental conduit. One example of the *Power of weak ties* is how people find out about jobs. Apart from advertisements and headhunters, the main source of information is not from close friends and family or existing employers, but via informal networks, friends of friends, and information randomly accessed from other sources.

Or take two communities threatened with disruption by major road projects. In one community, there were plenty of strong ties within monolithic organizations, like the church, a few large organizations, and the local council, but few links across the boundaries of the strong ties. Each group kept itself to itself. In the other community, there were few strong ties, but a multiplicity of weak ties between smaller clusters of interest groups, none of which was itself very strong or influential. In the first community, the monolithic groups each kicked up an enormous fuss but could not mobilize across a broad front; each group was ignored. In the second community, the weak ties gradually produced a groundswell of united, *ad hoc* protest that was successful in blocking the road project.

Silicon Valley epitomizes the power of weak ties. Here are a large number of independent, fiercely competitive businesses—it is far from a monolithic industrial structure, where there are a few large firms enjoying cosy relationships with each other—and yet there are ties between the firms and the people. Executives swap jobs frequently, they gossip in bars, they go to industry conventions, they engage in *ad hoc* collaboration, they play sport together, they share information. By way of contrast, Route 128 outside of Boston has the same industrial structure—many competing firms—but far fewer of the informal, weak ties; people are much more secretive and insulated from each other, and engage far less in a wide range of social contacts.[14]

The power of weak ties is closely related to another power law we examined in Chapter 5, Jared Diamond's principle of intermediate fragmentation, which states that you don't want excessive unity and you don't want excessive fragmentation; instead you want your human society or business to be broken up into a number of groups that compete with each other but that also maintain relatively free communication.

So what? The power of weak ties illuminates both the 80/20 principle and the way in which influence can be exerted across boundaries. If weak ties will do, you don't need strong ties. You don't need ownership. You don't even need control. You can achieve 80 percent of your objective with only 20 percent of the ammunition. In fact, strong ties may be less effective than weak ones, because strong ties encourage a sense of internal identity that removes the appetite for, or the ability to digest, a whole range of weak external ties.

Von Foerster's theorem

Derived from modern cybernetic theory, *Von Foerster's theorem* says that the more rigidly connected parts of a system are, the less influence they will have on the system as a whole. Each rigidly connected element of the system will be more 'alienated' from the whole system. Excessive control can therefore be counterproductive, and will certainly lower the potential value and cohesion of the whole system.

Why is it that organizations often find it more difficult to deal with sister companies than those where there is no ownership link? Why do executives often find their own weak ties—their own informal networks held together by nothing more than chance, vague mutual empathy, and the possibility of doing useful business together—more useful than the strong ties of organizational culture and common interest?

The 50/5 principle

The 50/5 principle is very useful and very snappy. Typically, 50 percent of a company's customers, products, components, and suppliers comprise a mere 5 percent or less of sales and profits. The profits may even be negative. Therefore, eliminate the low-volume customers, products, components and suppliers. Result: only a very small drop in sales, but a large reduction in complexity. Whether or not the numbers show it, an increase in real profits is almost inevitable.

For instance, in the early 1990s, Corning conducted 50/5 analysis at two plants producing ceramic substrates for auto exhaust systems, one in Greenville, Ohio, and the other in Kaiserslautern, Germany. The 50/5 principle worked. Out of 450 products made at Greenville, half produced 96.3 percent of total sales. The other 50 percent of products yielded just 3.7 percent. At the German plant, the bottom half of products produced only 2–5 percent of sales (depending on the time period analyzed). In both places, the bottom 50 percent made losses. They were eliminated, resulting in much simpler businesses, and, before long, a 25 percent reduction in engineering overhead.[15]

Because it is less radical and less threatening, it's often best to start with 50/5 analysis and then move on to 80/20 analysis.

Mendeleev's periodic table of elements

A methodology with strong similarities to the 80/20 principle was developed in 1869 by Dmitri Mendeleev (1834–1907), a great Russian chemist. Looking at chemical elements and trying to find their underlying unity, Mendeleev noticed that, as he said, 'the size of the atomic weight determines the nature of the elements.' Chemicals with similar properties have similar weights: manganese (to which Mendeleev's 1869 table assigned the relative atomic weight of 55) and iron (56), for example. *Mendeleev's table of elements* enabled him to successfully predict the properties of yet to be discovered elements.

Mendeleev's genius was to look for just one key variable that could explain different properties. This is similar to the concept of the 'vital few' as opposed to the 'trivial many,' except that in this case it was just one dimension that was vital. Recall the parallel with Isaac Newton's discovery that the force of gravity was proportional to the mass of bodies to the inverse of their distance from each other; in this case, just two variables were needed to make sense of movement everywhere.

Whether it is just one key variable, or two, or a few, the key to understanding anything seems to be to simplify it and identify a very small number of powerful causes. Not everything is susceptible to such discrete analysis, but many important things are. It is always worth trying to isolate between one and three important variables that may be causing most or all of what you are trying to explain: why profits have gone down, for example. If you can't explain most or all of the difference through fewer than four variables, it's probably time to give up analysis, and experiment instead.

Control theory

The 80/20 principle implies that we should reduce the number of things that we are trying to control. On the vital few things, however, it may be appropriate to exercise extremely tight control.

Control theory can help here. In physics and biology, control mechanisms regulate dynamic processes to achieve the controller's objectives. The basic idea is terribly simple: you monitor what is happening and if the system isn't doing exactly what you want it to do, you whack it back on course via repeated corrections (called 'negative feedback').

Many biological processes exemplify negative feedback. The population of a species rises. Food becomes scarce. The population then falls back to a sustainable level.

Thermostats and air conditioning are based on control theory. Attempts are now being made to use it to control complex systems such as irregular heartbeats, nerve impulses or artificial satellites. Whereas many attempts by humanity to exert control over the environment have proved disappointing, control theory, because it is part of the warp of nature itself, may prove to be an extremely effective tool. According to Ian Stewart:

> *In the future we may well use it [control theory] to control the flow of turbulent air past an aircraft wing, the population of codfish off the coast of Newfoundland, or the migration of locusts in North Africa. And we may use it to send supplies to our newly constructed Moonbase using only half the fuel now required.*[16]

Control theory can only work if it is possible to define the outputs you want precisely, and to measure them precisely too. Budgeting is a form of control mechanism, and the accounting systems that reduce the complex reality of business life to a few figures that can be measured and monitored are a good example of the value and limitations of control theory: the system of budget monitoring works quite well, but only if we all believe in the validity of the numbers presented. In reality, the information presented is necessarily distorted and only part of what should really be monitored, yet alternative systems have proved too complex and unworkable.

Therefore, if you are going to attempt to control something, you do need to think very carefully about the value of your measurements, and whether you will really be achieving the results you want. It's a boring cliché that what gets measured gets done. Yet the truly important things are usually the ones least measured and monitored, either because it is inherently impossible to do so, or because insufficient thought is given to what the measurement means. As an example of the latter, take a new product launch where the system monitors rates of trial by consumers. This is frequently done, yet it is the wrong measure: what matters is not trial, but repurchase rates after trial. Because repurchase is much more difficult to track, the wrong measure is often used—and the whole effort has no value.

Only use control theory if you are sure that you are going to measure the really important thing and that it can be measured precisely. Then:

◆ Define precisely what you are trying to do and how you will measure the output.

◆ Measure the output.
◆ Expect the output to differ from the plan.
◆ Correct the output to achieve the plan by applying feedback, or, if the market is telling you clearly to do so, change the plan.
◆ Keep repeating the process until the output is in line with the plan.

Fermat's principle of least time

The French mathematician Pierre de Fermat (1601–65) discovered that a ray of light traveling between two points will go the way that takes least time, not the shortest route. His mathematical proof of the *Principle of least time* led to the laws of reflection and refraction.

In going the quickest way, light is minimizing its scarcest resource: time. We can apply the same principle in business by thinking not about time, but about what is the scarcest resource. Affairs should be organized so that the scarcest resource is used most parsimoniously; things should be arranged for the convenience of the scarcest resource.

Reflect on what's really the scarcest resource in your business. If it is the time of a person, affairs should be arranged to make the most of this. There is a word for this, and it is not empowerment. The word is delegation.

Delegation is not always important, but when the scarce resource truly is a person nothing could be more crucial. The principle of delegation has, of course, been known for ages. Yet I have only ever come across one firm, among the hundreds I have known well, that effectively practices delegation.

The firm is Bain and Company, the eccentric firm of management consultants where I was briefly a partner some time ago. At Bain it was drilled into everyone: never do something that a cheaper or less experienced person could do.

This was a wonderful philosophy for a partner. All of life's tedious challenges, from buying a sandwich to choosing a suit, could be taken care of by juniors (I was going to write 'slaves') of one kind or another. Time billed by partners to clients—the scarcest and most valuable resource—was thereby maximized.

Trichotomy law

My final power law to help us achieve more with less is the **Trichotomy law** from mathematics. This states that every real number is either zero, or negative, or positive. This may seem trivial, but reflect on it in the context of value creation. We are back to 'less is more.'

In organizations and life generally, we tend to focus on the value that is created, ignoring the value that is subtracted. For example, the management hierarchy clearly has value in helping the top people achieve the organization's objectives. Or individual executives clearly add value in the course of their jobs. Or the organization benefits from having a sister division that is able to share some costs, such as a joint salesforce.

But this is not the end of the story. Every activity, unit, or person that adds value may also—and probably does—subtract value as well. The management hierarchy may demotivate people down the line, insulate them from thinking about the business themselves, or lead them to pay more attention to their bosses than to customers. Individual executives may have great strengths, but great weaknesses too, that require clean-up or damage limitation exercises all around them. The sister division may share the cost of the salesforce, but limit its effectiveness in selling your most profitable products, which may require a focus on a different sort of customer.

For every plus, there's likely to be a minus. What matters is the net result. We nearly always err in looking at the positive side and neglecting the negative.

One reason that we can achieve more with less is that we can decide not to do things that actually have negative value. Here we benefit twice: once because cost is removed, and again because we remove a negative effect that is greater than the positive effect.

The easiest, and often the best, way to add to personal or organizational effectiveness is simply to stop doing things that subtract net value. If you can't see what you routinely do—or what the organization routinely does—that subtracts value, ask your colleagues. Be prepared for a long list!

Summary

To create wealth, we must do more with less.

The 80/20 principle describes how the world works and shows us how to do more with less. The secret is to identify the powerful forces that have more than their fair share of impact. If they are forces that can help us, we should try to maximize them and ride them. If they are harmful forces, we must remove or avoid them.

Because roughly 80 percent of corporate assets, activity, and decisions lead to only 20 percent of profits or value, it follows that corporations should be much more selective and careful. They should own less, acquire less, divest more, focus on fewer stages of the value chain, and reduce their number of products, customers, suppliers, and employees.

The positive side is to find the 20 percent of activity that adds 80 percent of the value, and increase it.

The 80/20 principle also applies to individuals, and to their effectiveness in their careers and private lives alike.

Corporations and executives should seek to control far less, but to control the few things that really matter much more rigorously. Control theory can help here, but only if there is very careful thought about the objectives and measurements.

The best way to start to get more from less is to stop all activities that subtract more value than they add.

Action implications

◆ *Achieve more with less.* Make this your resolution every year, month, week, and day.
◆ *Start by applying the 50/5 principle.* Identify the least important or profitable half of the number of products, customers and suppliers that contribute only 5 percent of sales. Cut them.
◆ *Move on to the 80/20 principle.* Identify the 80 percent of products, customers, suppliers, and employees that contribute only 20 percent of value. Make them much more profitable or productive, if this is possible, or, if it isn't, remove them over time.

◆ *Focus all your energies on increasing the 20 percent of business*—whether defined by customers, products, or any other measure—*that contributes 80 percent of the value.* Try to sell more of the same or similar products, to the relevant customers or others who share similar characteristics.

◆ *Be extremely sparing in what you own,* the capital you use, the acquisitions you make, and the number of stages of value added in which you participate, and what you try to control. Over time, make your company more virtual and more focused on the minority of activity that delivers most of the value.

◆ *Make your firm as simple as possible.*

◆ *Develop skill in managing and exerting influence beyond your organization's boundaries.*

◆ *Identify the scarcest, most valuable resource in your organization,* and arrange everything else to make the best use of this scarce resource.

◆ *Think about and measure value subtracted as well as value added.* Identify whatever activities or links subtract more value than they add, and cut them. Stop any activity that you or others engage in, if the value subtracted is nearly as great as, or is greater than, the value added.

Notes

1   Diane Coyle (1997) *The Weightless World*, Capstone, Oxford.

2   The researchers are Bernardo Huberman and Lada Adamic; see the *New York Times*, June 21, 1999.

3   'Chaos theory explodes Hollywood hype,' *Independent on Sunday*, March 30, 1997.

4   Vilfredo Pareto (1896/97) *Cours d'Economique Politique*, Lausanne University. For a full explanation of Pareto's findings and how they can be used, see Richard Koch (1997, 1998) *The 80/20 Principle: the Secret of Achieving More with Less*, Nicholas Brealey, London.

5   With the possible exception of his contemporary, W. Edwards Deming.

6   See Chapter 1. Darwin's theory can be reduced to three observations: the struggle for existence among creatures resulting in the early death of most embryos and siblings (the insight from Malthus); the variations between and within species; and the inheritance of variation. Darwin then jumped to the

conclusion that the variations facilitated selection, since nature could reward variations that were most suited to the conditions of life. If he had started with the 80/20 principle, Darwin could immediately have hypothesized that a small minority of the most powerful variants would eventually populate most of their species; and that a minority of siblings would leave a majority of descendants. Thus two out of three of the planks of Darwin's theory are at least implicit in the 80/20 principle (only the inheritance point is not implied).

7   See footnote 4 above. Since you've bothered to read the endnote, here's the 80/20 insight into negotiating a pay hike. It's likely that about 80 percent of concessions will be made in the last 20 percent of negotiating time. So don't peak too early in your demands. If you start the meeting at 5.30 pm and you know your boss has to leave the office at 6.30, the critical moments will occur around 6.20. Try not to allow things to get resolved before then. If proposals are made before then, look unhappy and keep your own suggestions until near the time when the supervisor has to rush off.

8   I am drawing here on pioneering work undertaken by the Ashridge Strategic Management Centre, and in particular one of its directors, Marcus Alexander, for which I am most grateful.

9   Marcus Alexander (1997) 'Managing the boundaries of the organization,' *Long Range Planning*, October, 30 (5), pp 787–9.

10 As Marcus Alexander points out, 'virtual' can mean two different things. It can mean either lack of physical proximity, or, as here, lack of ownership. See the article from which my examples are taken and which contains many more: Marcus Alexander (1997) 'Getting to grips with the virtual organization,' *Long Range Planning*, February, 30 (1), pp 122–4.

11 Based on an unpublished paper by Marcus Alexander (1999) 'The value in corporate alliances,' draft prepared by the Ashridge Strategic Management Centre for the Singapore Chambers of Commerce.

12 See, for example, a study of 39 middle-sized German companies: Gunter Rommel (1996) *Simplicity Wins*, Harvard Business School Press, Cambridge, MA.

13 I have invented the concept of ROME, based on the 80/20 principle, and this is the first time that I have written about it.

ROME for any business segment may be defined as the percentage of total profit before interest and tax (PBIT) divided by the percentage of total management effort (ideally weighted by the cost of that management effort) going into looking after that business segment.

Thus we can express ROME arithmetically as:

$$\text{ROME} = \frac{\text{Percentage of Profit Before Interest and Tax}}{\text{Percentage of Management Effort}}$$

A ROME of more than 1.0 indicates a segment of above-average profitability, and the higher the number the better. A ROME of below 1.0, and especially one below 0.5, should lead to one of the following actions:

◆   a reduction in management effort, and/or
◆   an increase in profits, and/or
◆   withdrawal from the segment.

The hypothesis derived from the 80/20 principle is that segments taking approximately 20 percent of total management effort (those with the highest ROME) will account for 80 percent of PBIT. These segments should be expanded.

14 Jared Diamond, 'How to get rich.'
15 George Elliott, Ronald G Evans and Bruce Gardiner (1996) 'Managing cost: transatlantic lessons,' *Management Review*, June.
16 Ian Stewart (1989) *Does God Play Dice?*, Basil Blackwell/Penguin, Oxford/London.

11

On Punctuated Equilibrium, the Tipping Point, and Increasing Returns

First there is a mountain
Then there is no mountain
Then there is.

Zen proverb

Punctuated equilibrium

At the start of Chapter 4, we looked briefly at **Punctuated Equilibrium**, the theory that evolution consists of long periods of stability, punctuated by short periods of rapid transition. When a species evolves, conditions can be stable for several million years. Then there is a sudden leap, which takes place very quickly, and new species are formed. Evolution proceeds by a series of lurches, and everything is different.

Here we will explore this key power law in more detail, with particular stress on changes in technology and the way they follow the pattern of punctuated equilibrium.

The inventions of the nineteenth century, such as railroads, gas, electricity, and automobiles, marked such a punctuation point. Probably of roughly comparable importance in the later twentieth century were the surge in computing power, telecommunications, genetic engineering, and the internet.

Technological change is *the* main determinant of long-term growth everywhere. Besides opening up new possibilities and creating new needs, technology provides more for less; as well as driving up standards, it lowers costs dramatically. Indexed to 100 in 1930, the cost of air transport per mile and per passenger had fallen to around 17 by 1980. The charges for using satellites fell from 100 in 1980 to about 15 in 1990. The cost of a three-minute phone call from New York to London fell from around 100 in 1940 to about 2 today.

The most important technological changes are the 'general purpose' or 'enabling' technologies that transform economies and societies, that punctuate the equilibrium. The historical highlights of enabling technology include domestication of crops and animals, writing, bronze, iron, the water wheel, the windmill, the three-masted sailing ship, the printing press, automated textile machinery, the steam engine, electricity, the internal combustion engine, and the computer. Growth is very largely a function of the extent and speed with which enabling technologies are used, adapted, and spread.[1]

Recall the great Joseph Schumpeter's insistence that capitalism proceeds via 'creative destruction' brought about by technological change. When a punctuated point is reached, the implications can be exhilarating or scary, depending on which side of the divide you sit. If you are a traditional farmer, determined to follow the way that crops have been planted and grown for 1000 years, the arrival of genetically modified crops is a threat. If you are a leader in the new methods, and can triple productivity while also raising resistance to disease, there is a big opportunity.

The 'warning period' of embryonic pre-change

Punctuated equilibrium has some very intriguing and potentially profitable characteristics. Technology may lie latent or largely unexploited for several years, before suddenly taking off. The potter's wheel, for instance, was invented before 1500BC but only applied to spinning two and a half millennia later. Eye glasses were invented in the thirteenth century by the monk Roger Bacon (who died in 1294), but only exploited on a mass scale in modern times. Leonardo da Vinci drew prototype helicopters, and much else besides, but lacked enabling power mechanisms.

Capitalism's advance in the eighteenth century greatly shortened the period between invention and rapid deployment of new technologies. Steam power was applied within decades to every conceivable use: factories, steamships, railroad locomotives. Markets supplied incentives to introduce and diffuse new technologies.

Even today, however, new technologies do not take off immediately on invention. Instead, they spend years lurking in the profitless limbo of the enthusiast and the pioneer before exploding into mass markets: cell phones, video recorders and the internet being examples. The pattern is not simply stability/punctuation/new equilibrium, but rather stability for a long time, followed by a much shorter but significant period of embryonic change, followed by the punctuation and then rapid transition to the new equilibrium. Change does not happen out of a blue sky. There is a 'warning' or 'warm-up' or 'pre-punctuation gestation' period when the nature of the embryonic change is apparent to alert eyes and yet has not happened.

Plague theory

Plague theory charts how infectious diseases spread. It turns out the progress of any such disease, from the Black Death to AIDS to less serious infections, can be projected fairly accurately by calculating the proportion of the relevant population that has been infected at a few different time points (years, months, or days) and then extrapolating what will happen if the rate of growth of the disease, as measured by the ratio of the infected to the not yet infected, remains constant. It is reasonable to assume that the rate of infection will be roughly constant, if we adjust for the proportions of the population that, at any given time, are infected or healthy.

It is worth explaining the formula, because it is particularly useful in estimating how fast any new technology or new business method will spread. Assume that in month one, 100 people have the plague out of a population of one million; so 999,900 don't have the plague. By month two, 500 people have the plague and 999,500 don't. In the third month, 2500 are infected and 997,500 are healthy. With just three data points like this, there is a good chance that we can predict how many people will have the plague in months four to 48. By month four, some 12,300 are likely to have the plague and 987,700 to be uninfected. The formula for the projections is:

$$x = \frac{f}{1-f}$$

where:

x = the factor by which to measure growth

f = the percentage of the relevant population with the plague

And $1-f$ = the percentage of the population not infected with the plague.

The theory holds that the rate of growth in x will remain constant. What typically happens is that the proportion of the population affected by the plague will grow rapidly and at an accelerating rate, then will reach a point of inflection and slow down, and then decelerate rapidly: a typical S curve. The fast initial growth comes when the number of people who are infected and can therefore infect other people reaches a certain critical level. The rate of increase in infection slows down when most of those susceptible to infection have been infected and the disease runs out of new people to colonize.

The tipping point

If action can be taken to keep the disease below a certain critical level, then it may never reach the point of rapid acceleration, and the proportion of the population affected may end up being a fraction of what it would otherwise have been. The point of at which rapid acceleration begins is known as the **Tipping Point**, when the disease 'tips over' from being a low-level outbreak to a fully blown public health crisis. We can think of it also as the time that the disease acquires 'critical mass.'

The tipping point is therefore very similar to the time of 'punctuation' in punctuated equilibrium. The metaphor is more graphic. One moment you are pushing water uphill. Then you reach the tipping point and it's all downhill from there.

All new technologies that punctuate an existing equilibrium, and then replace it with another equilibrium, must pass their tipping point. If they don't, they will never become dominant or cause significant change to a business system. The same idea can be applied to any new product, fad or trend. A new social habit such as jogging, taking ecstasy, roller-blading, or following a new rock group, may initially make little headway. It may stay confined to a small subculture or area. Then it may begin to gather momentum. If it crosses an invisible line, it may never look back. If it doesn't cross the line, it will remain a small minority interest. That invisible line is the tipping point.[2]

The idea of the tipping point is enormously valuable if you are trying to launch a new technology or product, or trying to assess what impact someone else's innovation will have. The key point is that at no stage is effort or expense proportional to results. In the early stages, a terrific amount of cash and energy may be invested, with little apparent payoff. At this point, many pioneers cut their losses. Yet if the tipping point can be reached, it's all downhill from there. Sales and profits snowball, with relatively little incremental investment.

Microsoft's profits were tiny for its first 10 years. But once the tipping point was reached, around 1985, they exploded. So too for Federal Express. Beyond its tipping point, in the early 1980s, it was impossible not to compound profits like crazy.

The power of the unexpected

Peter Drucker points out that the easiest and simplest innovation opportunity lies in unexpected occurrences.[3] In the late 1940s, everyone knew that the only sensible use of computers was for advanced scientific work. IBM picked up an unexpected source of interest: certain large business firms, who were not then users of computers at all, indicated that they might want a machine that could run the payroll. IBM was much smaller than Univac—the UNIVAC (UNIVersal Automatic Computer) was the first general-purpose electronic digital computer designed for commercial use—but through providing a machine specifically for payroll applications, IBM overtook Univac within five years. Unexpected successes, shrewdly observed, can transform an industry and overturn competitors' rankings.

Drucker also demonstrates that unexpected *failures* may be just as fertile a source of insight as successes. The Ford Edsel was the best planned and designed car in automobile history. Yet it stunned the industry by being its biggest failure. Ford very wisely wanted to find out why: what was happening in the industry that ran counter to everyone's assumptions? The Ford people discovered that the prevalent segmentation of the industry by income groups was giving way to the new segmentation of lifestyles. Hence Ford developed the Mustang, a personality-based car that restored the company's fortunes.

The unexpected is often a tremendous clue to developments that are reaching their tipping point.

Crossing the chasm

In the 1950s, marketing theorists proposed a useful model of how new technologies and products are adopted—first by innovators, then by early adopters, then by the early majority of users, then by the late majority, and finally by the troglodyte laggards. An original and incredibly useful twist to this general model was proposed in the 1990s by Geoffrey Moore, a California-based high-tech marketing guru.

In his brilliant book *Crossing the Chasm*,[4] Moore points out that in the early days, it is easy for a new technology or product to sell itself, since the people who will try it will be 'innovators' who love technology or something new for its own sake. Typically, the founder of a high-tech company will be a 'techie,' an enthusiast and evangelist of the new way. For him selling to innovators is natural and involves no need to reframe the message.

But when it comes to selling to the early adopters, and still more to the early majority of users, there is a major barrier or gap to be crossed: what Moore calls the 'chasm.' The chasm is there because the mainstream market is not impressed by technology *per se*; if anything, the mainstream market is intimidated by technology. The mainstream market wants better performance, lower cost and all the boring normal purchase criteria; and it wants to be sure above all that the new product and technology is reliable, here to stay, and an integral part of the mainstream market, rather than a plaything for technophiles. This requires a quite different marketing and selling approach to that of the early days; technological enthusiasm becomes counterproductive; the message must be functional benefits and superior performance. This is the chasm that many incipient technologies and very young companies cannot cross. If they can't cross the chasm, they disappear into it, never to be heard of again.

Moore's model helps to explain why there is what I call the warning or warm-up period, during which the new technology is evident but not yet conquering the world. The model also gives the astute observer the tools with which to predict whether or not the new technology will make it, and some insight as to when. Unless the technology can sell itself to mainstream customers, which is very unlikely, it requires particularly careful product design and marketing to make it very user friendly. The technology will only cross the chasm if it is dressed up to look much less innovative and subversive than it really is. The technology needs the Trojan horse of mainstream marketing to persuade mainstream customers to let it into their lives. Unless and until this is evident, the new technology will stay in the wilderness. Cars must appear like horse-drawn carriages (hence 'horseless carriage'). Airplanes have to look like trains (with aisles and windows and an engine at the front), not like birds or bats. The PC has to resemble a typewriter. The internet has to seem like an extension of earlier software tools, linked to the well-established PC.

Exponential growth

Albert Einstein, asked what was the greatest force in the world, replied without hesitation: 'compound interest.'

It takes a genius to really understand the nature and impact of growth over a long period. Experiments have shown that even educated and highly numerate people tend to profoundly underestimate the impact of growth. For example, in one study[5] subjects were asked to estimate the necessary capacity for a tractor factory that starts in 1976 with a capacity of 1000 tractors and where demand increases at 6 percent a year: how many tractors, they were asked, would the factory have to make in 1990, 2020, 2050, and 2080? The typical answer showed a gradual and linear increase, and the estimates up to 1990 were pretty close to the real answer. But thereafter the correct answer shot up 'exponentially' while the estimates continued to show steady growth. By 2080, the typical respondent estimated that around 30,000 tractors would be needed, whereas the right answer was about 350,000, more than 10 times that level!

Or try this puzzle. One lily pad, covering 1 sq. ft, sits in a pond with an area of 130,000 sq. ft. After a week, there are two lily pads. After two weeks, four pads. Estimate how long it will take to cover the entire pond.

After 16 weeks, half the pond is covered. Now estimate again: how long before the whole pond is covered with lilies?

It has taken the lily 16 weeks to cover half the pond. Yet the right answer is that it will take only one more week to cover the whole, since the lily pads are doubling their domain every week—17 weeks in total.

Remember the fable about the Indian king who wanted to reward the inventor of chess? All the inventor wanted was a few grains of rice: one for the first square on the chessboard, two for the second, four for the third, and so on for all the squares. The king thought this was modest— until it was computed that for the last square alone, some 9,223,372,036,000,000,000 grains would be required: about 153 billion tons, or more than two and a half million big (60,000 tonne) cargo ships packed to the gunwhales with rice. This is because of 'exponential' growth, in this case the doubling of rice on every square.

What exactly is exponential growth?

An exponent is a number saying how many times something should be multiplied by itself. For example, if the exponent is 3, and the number is 4, then the expression 4^3 means $4 \times 4 \times 4$, which equals 64. In the mathematical expression y^2, 2 is the exponent, and it means $y \times y$.

How is exponential growth different from linear growth? In linear growth, something increases in size by the same *amount* at each step, not by the same *multiple*. If I start out owning $1,000, and increase it by $100 each year, after 10 years I will have doubled my money to $2,000. This is a linear increase, the same amount each year. But if I start with $1,000, and increase it by 10 percent each year, after 10 years I will have $2,594. This is exponential growth, a constant multiple (1.1) of growth each year. If I carried on for another 10 years, linear growth would give me a total of $3,000, but exponential growth $6,727.

Any market or business that grows at 10 percent or more for any significant stretch of time will have a far greater effect in terms of value creation than we would intuitively estimate. Some businesses—such as IBM or McDonald's in the period from 1950–85, or Microsoft in the 1990s—managed to grow at more than 15 percent per annum, producing fantastic increases in wealth. If you start with $100 and grow it by 15 percent per annum for 25 years, you end up with $3,292, nearly 33 times as much as you started with. A slightly higher gain in percentage terms ends up making a large difference.

For example, American stock picker William J O'Neil ran a fund for his classmates that started with $850 in 1961 and ended up with $51,653 in 1986, after all taxes had been paid.[6] Over 25 years, this is an average increase of 17.85 percent each year, producing a total that is 61 times the original stake. Thus, 15 percent per annum for 25 years produces 33 times the stake, but adding fewer than 3 percentage points to the growth rate, at just under 18 percent, produces a 61 times increase.

Exponential growth changes things qualitatively as well as quantitatively. For example, when an industry grows fast—Peter Drucker says when it grows by about 40 percent within 10 years—its structure changes, and new market leaders often come to the fore. Markets grow fast because of innovation, discontinuity, new products, new technologies, or new customers. Innovators, by definition, do things differently. The new way rarely fits the habits, ideas, procedures, and structures of established firms. Innovators may make hay for several years before traditional leaders counterattack, and then it may be too late.

Fibonacci's rabbits

Here's a fascinating puzzle about exponential growth. In 1220, Leonardo of Pisa, who was nick-named 'Fibonacci' 600 years later, constructed the following scenario. Start with a pair of rabbits. Then imagine that each pair gives birth to another pair one year, and a second pair the year after that. After that, they're too old to breed. How does the number of pairs of rabbits progress, and is there anything intriguing about this?

You can work it out if you want, but here's the answer. The number of pairs in each year is as follows:

1, 2, 3, 5, 8, 13, 21, 34, 55, 89, 144 ...

Can you spot anything weird about this?

There are, actually, two amazing things. One is that from the third number onwards, each following number is the sum of the two preceding numbers. The other is that each year's number is bigger than the year before's (after year three) by a virtually constant ratio, which soon becomes very close to 1.618. In other words, there is a constant rate of growth of just over 60 percent.

There are good mathematical explanations for **Fibonacci's rabbits**, for which, fortunately, we don't have space.[7] The rabbits do, however, illustrate the power of exponential growth, as well as the fact that even apparently reasonable growth like this can't go on too long. After 114 years of Fibonacci rabbit growth, the volume of rabbits would exceed the volume of the universe, and all humans would be dead, smothered beneath the bunny mass. Fur-fetched indeed!

Big Bang

A more extreme form of exponential growth was probably responsible for the start of the universe. Astronomer and physicists now generally accept the **Big Bang theory**, according to which the universe started at an unimaginably small size and then doubled in a split second 100 times, enough to make it the size of a small grapefruit. This period of 'inflation' or exponential growth then ended, and linear growth took over, with an expanding fireball creating the universe that we know today.

Creation of any sort involves exponential growth. The interesting lesson is that, with exponential growth, you don't need to start big. In fact, you can start extremely small. If the universe can start with something so small that we can't imagine it, and expand to its current equally unimaginable size, then the initial size of a new business is totally irrelevant. The key requirement is a period of exponential growth, followed by a longer period of linear growth.

Insights on growth

The greatest opportunities for creation and growth occur at times of punctuated equilibrium, or, if you prefer thinking of it this way, at and immediately after the tipping point.

Punctuations and tipping points don't occur without warning. There is always a period, sometimes quite long, of pre-punctuation warm-up, when the existing system shows signs of instability and the new system is silently building momentum. For new technologies or new products, the tipping point will not occur until the innovation can appeal to the mainstream market, which means that it must be sold on grounds of conventional benefits, and that the revolutionary nature of the change (if there is one) must be downplayed.

Periods of rapid change and high exponential growth do not, typically, last long. A new equilibrium with a new dominant technology and/or competitor is likely to be established before long. Periods of punctuation are therefore exciting and exhibit unusual uncertainty. The payoff from establishing a dominant position in this short time is therefore extraordinarily high. Dominance is more likely to come from skill in marketing and positioning than from superior technology itself.

Most innovations fail. To buck the trend they have to 'cross the chasm'—or pass the tipping point—to reach the mainstream market. Acceleration is the key. Unless a new product or technology is accelerating, it's unlikely to make it.

Say's law of economic arbitrage

In 1803, French economist Jean-Baptiste Say (1767–1832) produced a remarkably modern work titled *A Treatise on Political Economy*. Thomas Jefferson pronounced it:

> *A very superior work ... his arrangement is luminous, ideas clear, style perspicuous, and the whole work brought within half the volume of [Adam] Smith's work.*[8]

It contained many surprising innovations, including coining the word 'entrepreneur,' and, in the same sentence, the first theory of economic arbitrage:

> *The entrepreneur shifts economic resources out of an area of lower productivity into an area of higher productivity and yield.*

Long before the notion of return on capital was promulgated, Say identified one of the most important engines of economic creation and progress. Resources are essentially finite, so growth depends not so much on the discovery and exploitation of natural resources, as on making each unit of resource go further. This is partly a function of better technology and methods, but also the entrepreneur's skill in moving resources to where they can be most productive.

Freud's reality principle

In 1900, Sigmund Freud (1856–1939) published *The Interpretation of Dreams* and founded the new science of psychoanalysis. One of his key concepts was the **Reality principle**, which says that what stops us from using other people for our own ends is that they are trying to do the same to us. Confronted by reality, we have to accommodate other people's needs and the demands of the outside world in order to obtain any satisfaction for our instincts.

Freud's concept is clearly valid, but a rather different spin was put on the same idea by his contemporary, dramatist George Bernard Shaw:

> *The reasonable man adapts himself to the world [in line with Freud's reality principle]: the unreasonable man persists in trying to adapt the world to himself. Therefore all progress depends upon the unreasonable man.*

Creation and entrepreneurship require the supply of new ideas, new methods, and unreasonable approaches. In insisting that automobiles should be bought by working men, and not just by the rich, was Henry Ford being reasonable or unreasonable? He certainly was not following demand, since there was no demand for cars except from the rich. Ford refused to accept the world as it was; he persisted in trying to adapt the world to his vision. By using the assembly line and maximum standardization, Ford cut the cost of a Model T from $850 in 1908 to $300 in 1922, and succeeded in his mission of 'democratizing the automobile.'

The successful entrepreneur

The *Book of Genesis* and the theory of the Big Bang agree on one point: there was only one genuine creation in the whole of history. Thereafter, progress means rearranging the pieces. There is nothing new under the sun.

Far from being a dismal view, this should be inspiring. *All* that is needed to add to human wealth is to take a given set of resources and shift them from areas of low productivity to areas of high productivity.

All economic progress rests on this economic arbitrage of this type. This is good news. Arbitrage is easier than creation. Everyone should be capable of thinking of something that could benefit from economic arbitrage, of identifying resources that could be used more effectively.

True entrepreneurs don't expect market research to tell them what to do. They have a vision of how to do something better and different, they work out how to do more with less, they shift resources from low- to high-value uses, and they are persistent and unreasonable until the world has conceded their point.

The law of diminishing returns

One of the most influential and long-running ideas about how markets and businesses operate is the *Law of diminishing returns*, developed around 1767 by French economist Robert-Jacques Turgot.

This law says that after a point, increases in effort or investment result in diminishing returns; that is, the incremental value declines. To a hungry woman, a loaf of bread has high value. The second loaf has less. The tenth may have very little value. If you hire additional peasants to till the same plot of land, beyond a certain point diminishing returns set in.

A century later, the British classical economists, led by Alfred Marshall, extended this idea to markets and firms. Products or companies that lead a market run into diminishing returns. The value of being bigger in business—having a larger market share, a bigger factory, a larger product range—reaches a peak and then declines. Again, this may sound like common sense.

But the classical economists went further still. They claimed that a predictable equilibrium of prices and market shares would be reached, and that perfect competition and diminishing returns should eventually operate to ensure that super-normal returns would be impossible. This theory justified government regulation of markets: if high returns were being made, it could only be because monopolists were rigging the market and obstructing perfect competition.

An attack on microeconomics by heretical consultants

For almost a century, Marshall and his school dominated economic thinking. It was left to a few business mavericks, such as Bruce Henderson and his Boston Consulting Group (BCG), founded in the early 1960s, to challenge the consensus. BCG demonstrated that:

◆ Firms typically do not have equal costs. The costs of the market leader are usually significantly lower than those of followers.

◆ Costs and prices do not reach a static equilibrium. In competitive markets, costs and prices continue to come down for ever. Cost and price decreases are particularly characteristic of high-growth markets.

◆ A firm with high relative market share should, and usually does, have both higher return on capital and intrinsic competitive advantages, compared to other firms in the same market. Far from being subject to diminishing returns, high market share makes it possible to rein-force competitive advantage, by providing better products and services at lower prices, while still earning higher returns than competitors. High market share can lead to a virtuous circle, which further com-pounds advantage for the leader.

Moore's law

At about the same time that BCG was publishing its ideas, Gordon Moore, the co-founder of Fairchild Semiconductor in 1957 and Intel in 1968, promulgated his law. In its original form, in 1965, Moore claimed that computing capacity would double every year at no extra cost, as semiconductor density doubles. He explains:

> I looked at the first few integrated circuits that Fairchild [produced] ... and just happened to see that we had about doubled the number of components on an integrated circuit every year. So I blindly extrapolated that for 10 years, from about 60 to about 60,000 circuits on a chip—a long extrapo-lation—and it was amazingly precise.[9]

Moore's law was updated in 1975 to say that the number of components on the chip would double every two years, and that has also proved pretty accurate. In 1999, Moore predicted that the gradient would change again and that 'it will double every four or five years for quite a long while.'

Moore's law was consistent with a more general power law enunciated in the late 1960s, BCG's experience curve, which said that costs come down by 20–30 percent every time the accumulated production of an item doubles. The change in the slope of Moore's law—where initially value doubled every year, then every two years—just reflects the change from hyper-growth to high growth, or an increase in the months needed for industry production to double.

Some industry observers began to suggest that the 'IT economy' was different from the rest of the economy, because it did not appear to be subject to diminishing returns. BCG, however, had an intellectual framework suggesting that the world of IT was just a faster-growth version of the whole economy. BCG's views did not, however, reach their tipping point; they remained largely ignored.

BCG lacked two things. It did not have the professional academic credentials to be taken seriously by economists. Nor did it find a snappy name to encapsulate its new version of economics.

The law of increasing returns

Then, around 1980, along came W Brian Arthur, an economist from Northern Ireland working in the United States, who was heavily influenced by the ideas of chaos.

Brian Arthur had both professional credentials and the bright idea of branding his thinking as 'increasing returns.' He also had the good sense not to take on the economic establishment head-on. His **Law of Increasing Returns**, he said, should be seen as a supplement to Marshall's ideas about diminishing returns, not as a replacement for them. Alfred Marshall, brilliant chap, ideas well suited to their day, still relevant to smokestack industry today: that's Arthur's line. But the new economy, especially the high-technology world, now that's a different matter. This brave new world is subject to increasing returns.

Brian Arthur makes his case persuasively and vigorously.[10] What if products and businesses that got ahead thereby got further ahead? Arthur gives the example of the market for operating systems in PCs. In the early 1980s there were three contenders, all in with a good chance: CP/M, Microsoft's DOS, and the Apple Macintosh system. CP/M had the advantage of being first. The Mac was probably the best, and certainly the easiest to use.

But operating systems for PCs exhibit increasing returns. If one system gets ahead, more hardware manufacturers and software developers will adopt it, causing it to get further ahead. The key event happened in 1980 when IBM gave Microsoft an exclusive deal to write the operating system for the IBM PC. Although the latter was not a great machine, the growing base of DOS/IBM users attracted independent software houses, such as Lotus, to write for DOS. Once DOS/IBM had established a clear lead, it was bound to get further ahead, because the costs to switch to another system were too high. Microsoft then benefited from economies of scale, being able to spread its costs over a larger user base than competitors, thus enabling Microsoft both to enjoy fatter margins and also to spend more to improve its system.

Increasing returns, says Arthur, are characteristic whenever markets have the following attributes:

◆ *High up-front costs*, especially in R&D rather than production. The first sale of Windows cost Microsoft $50 million; the second $3. These economics make leadership extremely valuable and difficult to challenge.
◆ *Network effects*. Many high-tech products must be compatible with a network of users. Therefore a popular product or system is likely to become the standard. Also, network economics are different from traditional economics. (Network effects are so important that I'll expand on them shortly.)
◆ *Customer groove-in*. High-tech products are difficult to use, and also have several generations of product. Users have to invest in training. This training 'grooves' or locks customers into the leading product.

Metcalfe's law

Networks comprise an increasingly important part of our world, and they have their own peculiar economic characteristics. This is easiest to see in products like the telephone, fax, a PC operating system, the FedEx courier system, or the internet. One phone or fax or email address is useless. Two have some value. Thereafter, any increase in the network size has a more than proportionate increase in value to each user.

Bob Metcalfe, the inventor of Ethernet, a local networking technology, noticed that small-scale networks were not viable, but that putting together small local networks sharply multiplied their value. In 1980 *Metcalfe's law* was born: the value of a network equals n squared ($n \times n$), where n is the number of people in the network. Thus a 10-person network is worth 100, but a 20-person network is worth 400: you double the network and quadruple its value. A linear increase in membership means an exponential (to be more precise, geometric) increase in value.

Network economics therefore exhibit an extreme form of increasing returns, both for all members of a network and for leading suppliers to the network. An expanding network becomes a self-reinforcing virtuous circle. Each new member increases the network's value, which in turn attracts new members. Indeed, network members are unpaid but well-rewarded evangelists of the network. You mean you don't have an email address yet?

Networks typically spend quite a while reaching their tipping point, and then there is no stopping them. For 20 years, fax machines struggled to reach their tipping point. Then, from about 1985, almost everyone was installing them.

Monopoly is desirable in a network. Who wants a separate airline for each route? Or three competing PC operating systems?

With networks, value comes from openness and from proliferation. Traditionally, value comes from having a closed, proprietary system and from scarcity. No more. Ask Apple whether keeping the Mac system proprietary was such a smart idea. Ask banks whether cash machines should be proprietary or shared. The more networks ally with other networks, the more valuable they become. The gain in coverage and value creation far exceeds the loss in the exclusivity of value capture.

Economist Paul Krugman observes that 'in the Network Economy, supply curves slope down instead of up and demand curves slope up instead of down.' The more you have, the more you want: the exact opposite of diminishing utility. The more we make, the easier and cheaper it becomes to make more. This is the beauty of network economics. It is both deflationary, in that prices come down for ever, and expansionary, in that more and more useful things are created and used.

Is there a 'new economy' and a 'new paradigm'?

The possibilities of networks in general and the internet in particular have led some observers to claim that leading competitors can come close to generating virtually infinite returns. One reason is the sheer low cost of internet transactions: it costs a traditional travel agent $8 to process a typical airline ticket versus just $1 on the Web; a typical bank transaction costs $1 off the Web but as little as 1¢ on it.

As the marginal value of a network increases with scale, so the average cost of software declines, since marginal cost is almost zero. Non-network businesses may have high fixed costs, but the marginal cost of meeting customers' demands never falls near to zero: sales, marketing and customer service are all expensive operations. By contrast, networks and in particular the internet may add customers and sales at negligible extra cost.

Not only this. Traditionally, there has been a trade-off between high-volume standard business and customized business. The latter has required extra expense, and thus has only been viable if customers pay more. But, post-internet, the cost of customization can be trivial, and if customization greatly increases volumes, it could actually lower average cost.

The internet also separates information flows from physical flows. Take a store: a supermarket or a book store. A store is both a physical entity—a warehouse—and a source of information to the shopper—what is on the shelves is what is available, and it may be inspected. But the internet separates the two. Amazon.com initially supplied the information without involvement with physical flows. It could therefore have huge stock with zero inventory, escaping the traditional trade-off between cost and choice.

Web economics also create the possibility that *consumers* of information also become unpaid *producers* of information, as when Amazon.com users add book reviews to the site.

Note finally that the cost of cross-selling different products decreases dramatically with the internet. If you are selling a flight ticket, you can very easily and cheaply sell hotel rooms, travel insurance, car hire, and many other services. The value of a loyal customer base can hardly be exaggerated.

But internet and network economics generally do not offer a bonanza for everyone: though the amount of new value created can be enormous, most of its goes to a few players in the industry. There is a further reinforcement of the normal tendency for returns to be distributed asymmetrically, because of the emergence of 'sweet spots' in the total industry chain.

The theory of industry sweet spots

The internet's separation of physical flows from information flows makes vertical integration unnecessary and tends to divide industries into a large number of 'layers,' separate stages of the value-added chain where independent firms specialize. But whereas traditionally it was an industry or segment leader that made high returns, now it is the leader in *certain layers only*, probably just one or two layers, who will make high returns— and everyone else in the industry, including leaders in the non-favored layers, may struggle to cover the cost of capital. The Boston Consulting Group calls the favored layers **Sweet spots**.

An excellent instance where the sweet spot in an industry takes a quite disproportionate share of the industry's total profits is the case of Microsoft in PC operating systems. Microsoft's near-monopoly of the software layer gives it a very large share of total industry profits, despite Windows comprising only 2 percent of the total industry cost structure. The actual production of PCs comprises 75 percent of the industry cost structure and capital employed, but only a small percentage of total industry profits.

The race to establish dominant standards in sweet spots

Establishing competitive advantage in the new environment requires recognition of the strategic layers in an industry—the sweet spots—then their domination, if necessary in alliance with another powerful industry player. To establish dominance in the sweet spot, in turn, requires establishing a dominant standard. Thereafter, it requires careful 'orchestration' of players in the other industry layers: the suppliers and users of the sweet spot products. To stop the orchestrated becoming powerful, you have to divide and rule, and ensure that no one else can supply a differentiated and valuable product. Otherwise, the orchestrated will bite back.

This, after all, is what happened when Microsoft was orchestrated by IBM in the 1980s, when IBM outsourced the design of its operating system software to Microsoft. Later, the balance of power shifted decisively, because Microsoft's market size and value increased faster than IBM's (because of open architecture, which allowed Microsoft to supply IBM's competitors), and because there was no attractive alternative to Microsoft's products.

Competitive advantage, based on standards, not cost, may be temporary

Because competitive advantage becomes based more on dominant standards than low costs, it may become more difficult to sustain. There is always the risk that an innovative competitor may find the next sweet spot in the system. Microsoft can't rest on its laurels. Netscape and its friends are promoting network computing, where any PC operating system, including Windows, is subordinated to a new high-value strategic layer controlled by a Java-enabled browser. To defend its dominant position, Microsoft scrambled to incorporate browser technology into its operating system.

So what has changed in the 'new economy'?

The number and diversity of competitive segments have increased, as vertical integration falls apart. There are more layers. Dominance still offers very high returns—in fact, higher than ever, because more customer value can be added at lower incremental cost. But choosing the part of the value chain to dominate—ideally the sweetest spot—is critical. Building this dominance is more a matter of establishing dominant standards than of having the lowest costs. Defending the dominance requires orchestration of players in other industry layers, so that no single orchestrated player can become differentiated and skillful enough to turn the tables and become the new orchestrator, by providing a new source of customer value and a new dominant standard.

Networks and particularly the internet raise the stakes: the winner takes most of the spoils, not only—as has always been true—within its own sphere of activity, but also now within the total industry, including areas where it has no assets employed. The basis of competitive advantage shifts from pure scale and the cost advantage it brings towards the scale enjoyed by having better products and dominant standards, and the opportunity to capture a very high share of industry value added from a small layer within the total industry. To sustain extraordinary returns requires eternal innovation.

In short, the system dynamics have become richer, the inequality in returns has become greater, and choosing where to compete has become even more important; but leadership remains crucial. From a macroeconomic viewpoint, a long period of high growth—while the new enabling technologies become fully used—may be possible; industrial structures and leadership may be transformed; returns on capital may rise; and temporary monopolies may become essential for the public good. But the 'new paradigm' is not really new—it just represents one more punctuation point within mankind's long history of economic punctuations—and the ride will be as bumpy as ever.

Are there two economies and two sets of economics?

Brian Arthur argues that there are really two different economies, and that different economic and management theory is applicable to each:

> We can usefully think of two economic régimes or worlds: a bulk-production world yielding products that essentially are congealed resources with a little knowledge, and operating according to Marshall's principles of diminishing returns, and a knowledge-based part of the economy yielding products that essentially are congealed knowledge with a little resources and operating under increasing returns…
>
> Because the two worlds of business—processing bulk goods and crafting knowledge into products—differ in their underlying economics, it follows that they differ in their character of competition and their culture of management. It is a mistake to think that what works in one world is appropriate for the other.[11]

In an interesting thesis, Arthur goes on to ask why the new management ideology of flat hierarchies, missions, flexible strategies, reengineering, and 're-everything' have emerged. His answer is that they are not fads, but correspond to the high-tech world of constant reinvention; equally, that hierarchies and old-style management are appropriate to smokestack industry:

> Marshall's world tends to be one that favors hierarchy, planning, and controls. Above all, it is a world of optimization.

Arthur admits that there is a middle ground between the old world and the new. He asks where service industries such as insurance, restaurants, and banking belong. His answer is that they have a foot in each camp. On the one hand, most services are low tech, consist of 'processing,' and are subject to regional limits on demand—all characteristics of the diminishing-returns economy. On the other hand, most services can be branded and are subject to network effects—McDonald's or Motel 6 franchises attract more than their fair share of custom because the brand is well known and reliable. And, over time, services are moving to the new economy. 'In services,' he says, 'everything is going software.' Information is key, and is now processed more by software than by people. So 'service providers become hitched into software networks, regional limitations weaken, and user-base network effects kick in.'

Is Brian Arthur right? His examples of the new world are spot-on. Yet is there really an 'old' economy that fits Marshall's diminishing-returns economics? Should we not rather think of a 'standard' economy and a 'new' economy, both subject to increasing-returns economics, but to differing degrees?

Do Marshall's economics work at all?

The whole of microeconomics is an impressive intellectual edifice, constructed with mathematical elegance and, within the terms of its system, totally consistent and coherent. The problem is that it cannot be empirically observed. It does not correspond to the real world, not even to Marshall's world. That, too, was a time of high tech: the technological innovations of steam, railways, electricity, gas, and motorcars were at least as transforming as our own high-tech industries, and subject to network effects and increasing returns.

What was the Ford Motor Company, if not the Microsoft of its day? Every time Ford made a new Model T, in increasing quantities, the cost of each unit went down. Every time the cost went down, more people could buy one. Every time more Model Ts were bought, the cost of making the next went down. Every buyer demanded new roads, motels, and roadside restaurants, and mobilized support for initiatives that would make the motorcar yet more popular. When government began to build freeways, this network effect accelerated the car's spread. Plainly, autos manifested increasing rather than decreasing returns.

But the problem with Marshallian economics goes deeper than a failure to notice networks or high tech. If classical economics has any validity at all, it is in the pre-industrial world, related to commodities and to agriculture before it was mechanized. Gold, silver, iron; potatoes, wheat, cotton—these may be subject to diminishing returns. (Remember that the law of diminishing returns was developed in pre-industrial France.) When supply goes up, price goes down, but costs do not, so increasing supply is bad for producers. And as long as all producers use the same means of production, and there are no economies of scale or experience, then markets will behave as Marshall predicted and come to an equilibrium point where capital cannot earn any differential super-profits.

Classical economics does not work, however, when any of the following conditions applies:

◆ There are economies of scale or experience, so that the largest producer has lower costs, and increased supply lowers prices, resulting in higher demand, and still lower prices, in a virtuous circle that can go on for ever.
◆ There are differences of technology used, so that one technology may come to have lower costs than another—another reason that competitors may have different costs.
◆ Any competitor finds a better or cheaper way to do something, again contributing to unequal margins between suppliers.
◆ Goods cease to be commodities, because one manufacturer adds extra value by means of branding, product differentiation, or better service.
◆ There are high fixed costs in production.
◆ There are substantial barriers to entry for new suppliers.
◆ There are network effects.
◆ Human ingenuity can make more out of less, so that the material costs become a very small part of the total (as with the silicon chip, made with sand).
◆ Resources—such as information—are enhanced rather than used up as production expands.

Most business since the nineteenth century has had at least one of these characteristics and therefore requires dynamic economic analysis.

There are only degrees to which classical economics is inapplicable and misleading, and degrees to which the 'new economics' is applicable and helpful.

To those of us brought up on Bruce Henderson's insights into the value of market share and the astronomical value of 'star' businesses—the leaders in high-growth markets—the 'new economics' is not so new.

Different management styles for different types of business?

A final note on Brian Arthur's hypothesis that the 'new economy' requires new management structures such as flat hierarchies and 're-everything.' This hypothesis also is not new. It is essentially a restatement and updating of the argument in Tom Burns and G M Stalker's classic book *The Management of Innovation*, published in 1961. Stuart Crainer says that the book 'identified the 'organic' organization characterized by net-works, shared vision and values, and teamworking.'[12] Burns and Stalker also argued that it was precisely high-tech and high-growth organizations that required the new management approach; and that slower-growth firms in more predictable environments were more suited to command-and-control methods.

But are Arthur, and Burns and Stalker, correct? If there are, in fact, only gradations of the 'new economy,' and nearly all businesses in fact belong, to a greater or lesser extent, to that 'new economy'—which is not really new, but just a better description of the world after the Industrial Revolution—then different types of company may not need very different styles of management. Don't all firms need 'observation, positioning, flattened organizations, missions, teams, and cunning'?

Equally, don't *all* organizations need hierarchy, and have it? Isn't Microsoft a meritocratic dictatorship run by Bill Gates, that would be worth much less otherwise? What creates value is insight plus hierarchy.

Microsoft is an ideal model for any type of business: a dictatorship of ends, a meritocracy of execution, and a collegiate, 'democratic' style that respects intelligence and insight at every level, so long as it does not challenge the basic strategy.

Insights on networks, the 'new economy,' and increasing returns

Traditional microeconomics always was a poor guide to the real economy of business.

Whenever products cease to be undifferentiated commodities, whenever one competitor has and can retain a cost or product/service advantage over other suppliers, whenever brands or standards are important, whenever there are high fixed costs in a business and low incremental costs, whenever there are substantial barriers stopping new players coming in to a market—whenever one or more of these conditions holds, we are in a dynamic economy where equilibrium is elusive and advantage goes to the leader, who may enjoy and compound high returns.

Even 100 years ago, most businesses lived in such a dynamic economy. Today, the vast majority do.

Some things have changed. While for a long time nearly all businesses have been part of the dynamic economy, the extent to which they have been subject to the laws of dynamism has been steadily increasing. Fixed costs have risen. The cost of overhead, in the form of highly qualified and expensive professionals, has steadily increased; the cost of materials and unskilled labor has steadily declined. Incremental costs have declined. Technology and knowhow have become increasingly important. The risks and returns of business have increased. The value of leadership in high-growth markets, which was always substantial, has increased still further.

The shift towards winner-take-most economics has been most pronounced in network businesses, and, at the extreme, in electronic business. Here the stakes are raised. The few winners may make fantastic returns, grabbing most of the industry value added despite participating in only a thin strategic layer, the industry sweet spot. The many losers will be stuck in their cash traps.

Two long-standing rules of business strategy have only become more important. Do whatever is necessary to move ahead of competitors. And cut your losses when someone else has reached that point. To these we may add two new rules: Identify and dominate the industry sweet spots, by establishing new standards there, orchestrating others to do the donkey work in the bulk of the industry. And defend the dominance by dividing and ruling the orchestrated, and by continual innovation to find the next industry sweet spot.

Summary

Technology change drives growth and prosperity. But technology change is lumpy: we lurch from one long period of equilibrium to another via short, sharp periods of punctuation. Despite this apparent unpredictability, technological transformations follow a predictable pattern. Any change that will become significant gives plenty of warning. It accelerates as it approaches its tipping point, beyond which it becomes much easier and more rapid.

The easiest and best clue to transformation is unexpected occurrences—unexpected successes and, equally, unexpected failures. These always present huge, but usually neglected, opportunities.

Rapid market growth usually overturns established company pecking orders.

The value of a business that can establish new market leadership and sustain high growth for at least 10–15 years is usually greatly underestimated.

Entropy can be defeated by entrepreneurship. Everyone can be a successful entrepreneur, if they can spot an under-utilized resource and move it to a higher-value context.

Virtually all business is subject to increasing returns. Network businesses exemplify this to an extreme. But although returns increase for the total market, and all customers benefit, the benefit to firms is very heavily skewed towards the few leaders in the industry sweet spots. Most players in high-growth businesses will be losers, just like in any other business. The only difference is that the stakes are higher.

Action implications

◆ *Identify sweet spots in emerging networks and dominate them by creating a new standard of value.* Find the best possible ally or allies and strike a deal with them before anyone else. Confine your part in the industry to the chosen sweet spots and orchestrate suppliers carefully. Ensure that no supplier can develop its own distinctive standard on which you become dependent.

◆ *Don't play in network markets unless you can win,* or unless you have a fair chance of winning and are using other people's money to punt with.

◆ *Find a high-growth market that you can dominate,* even if it isn't a network market. Start by identifying a technology or superior way of doing business that is approaching, but has not yet reached, its tipping point. Become the best exponent of the new approach.

◆ *Find an undervalued resource and apply it to a new market.*

◆ *Cut your losses if you can't overhaul the market leader,* especially in markets with high fixed costs and low incremental costs.

Notes

1 See Martin Wolf (1999) 'Putting the paradigm to the test,' *Financial Times*, 10 November. The cost indices quoted are derived from OECD research by Professor Richard Lipsey of the Simon Fraser University, Canada.

2 Malcolm Gladwell (1996) 'The tipping point,' *New Yorker*, June.

3 See his classic article: Peter F Drucker (1985) 'The discipline of innovation,' *Harvard Business Review*, May–June, reprinted in November–December 1998.

4 Geoffrey Moore (1991) *Crossing the Chasm: Marketing and Selling Technology Products to Mainstream Customers*, Capstone/HarperBusiness, Oxford/New York.

5 See Dietrich Dörner (1996) *The Logic of Failure: Why Things Go Wrong and What We Can Do to Make Them Right*, Metropolitan Books, New York. Original (1989) published in Germany in 1989 under the title *Die Logik des Misslingens* by Rowohlt Verlag.

6 William J O'Neil (1991) *How to Make Money in Stocks*, McGraw-Hill, New York, p. 132.

7   The mathematically inclined can visit Peter M Higgins (1998) *Mathematics for the Curious*, Oxford University Press, Oxford.

8   Thomas Jefferson in a latter to Joseph Milligan, April 6, 1816. This is a splendid article and I have used it in my account.

9   Gordon Moore reported in John Naughton (1999) 'No goodbyes for world's Mr Chips,' *Observer*, 8 August.

10 The best short summary is W Brian Arthur (1996) 'Increasing returns and the new world of business,' *Harvard Business Review*, July–August.

11 Ibid., p. 103.

12 See Stuart Crainer (1998) *The Ultimate Business Guru Book*, Capstone, Oxford, p. 271.

12

On the Paradox of Enrichment, Entropy, and Unintended Consequences

Nothing fails like success.

Richard Pascale[1]

The power laws of caution

In just 13 years between 1970 and 1983, one third of the 1970 Fortune 500 top US firms vanished into the corporate Bermuda triangle. Few of them actually went bust. Most were taken over or merged with other companies. But still, it's a remarkable attrition rate.

In 1982, the most successful business book of all time was published: *In Search of Excellence* by Tom Peters and Bob Waterman. Two years later, *Business Week* ran a cover story under the strapline 'Oops,' gleefully chronicling the fall from grace of many of the 75 'excellent' companies. Later on, one of the most apparently impregnable and super-successful of the excellent companies, IBM, nearly went bust. Yet Peters and Waterman had been careful in their selections.

The average life expectancy of a multinational company, according to Arie de Geus, a former Shell executive, is between 40 and 50 years.[2] A study of firms of all sizes, covering Japan and most European countries, showed an average life expectancy of just 12.5 years.[3]

These three examples show what perhaps we all sense anyway: that it's difficult to sustain success, and that even very successful corporations suffer from the occupational hazard of all corporations—they live and die by the market and competition; and competition comprises not just that for customers, but also for corporate control via takeovers. No other major institution is as exposed to failure as is the corporation.

And so it should be. If corporations were not exposed, we would not enjoy high living standards. And the fruits of sustained success are so high—especially for investors and top executives—that it should be difficult to keep ahead.

Can systems thinking help us here? The answer is a qualified yes. There is no simple or overarching recommendation. Instead, we have to piece together insights from a number of 'power laws of caution.' In doing so, we learn to treat success with caution and humility.

There are three main power laws of caution:

◆ *The Paradox of Enrichment*.
◆ *The Law of Entropy*.
◆ *The Law of Unintended Consequences*.

The paradox of enrichment

Studies in ecology have confirmed that the number of predators and prey tend to oscillate together in fairly regular cycles. For example, the Hudson's Bay Company has kept records of the number of lynxes (predator) and hares (prey) since 1850, and graphs of these records show remarkable symmetry: they move up and down together, after a brief time lag.[4]

The prey depend on the predators quite as much as the other way round: without the predators, the prey will become too numerous and starve. Both populations benefit from what we may call a 'swinging cycle,' where the numbers swing up and down but without reaching unsustainable peaks or troughs—the cycle moves around a central point, or 'swinging equilibrium.'

Now here comes the paradox of enrichment: if some apparently benign environmental change allows the population of the prey to go up substantially, this may, strangely, be bad for both predators and prey. A large increase in prey leads to an even larger increase in predators, who before long find that they don't have enough prey to eat. The number of predators zooms up initially and then plummets; and the number of prey follows the same pattern—deprived of predators, there are soon too many prey, who can't find enough to eat. So a stable cycle, our swinging equilibrium, turns into an unstable cycle. The unstable cycle may end in disaster for both prey and predators, if one of the low cycles goes too far and the entire population of prey or predators is wiped out.

Because nature goes in cycles, the paradox of enrichment means that you really can have too much of a good thing. Take another example: trees are sprayed with insecticide, which kills harmful insects. But slightly too much is sprayed on the leaves, and then it rains. The excess insecticide is washed from the leaves to the ground, where it kills the insects' predators. Result: more insecticide leads to more insects.

The paradox of enrichment has a clear parallel in the classical economic theory that when a market is very profitable, it will attract new entrants, and profit will be driven back down to zero as a new equilibrium is found. From having too few firms in a market, before long there are too many. If you have read sequentially through this book, you will know that I'm not a great fan of classical economics, since it very rarely corresponds to real-world markets. Equilibrium rarely happens. But what happens in markets is that there are cycles, which we may divide into three types: swinging equilibrium, virtuous cycles, and vicious cycles.

Swinging equilibrium

Swinging equilibrium is the closest to the classical economists' dream. Equilibrium almost never happens, but the system swings up and down in predictable and functional ways just as control theory says it should. This is the world of Canadian hares and lynxes, of a thermostat controlling room temperature, or of the stock market's bear and bull oscillations (although in the latter case, there is a secular trend upward).

In my experience, only small parts of the business world follow this pattern. Commodity prices make a good example. Markets are only like this if they have very low barriers to entry and exit, undifferentiated products and services, and no possible advantages from scale, technology or doing things more cleverly than others. Some markets come pretty close to this, and the pickings are almost as meager as classical theory predicts: one can think of the market for cheap bed-and-breakfast accommodation, for unskilled labor, or, in poor countries, the selling of firewood or watermelons by the roadside. Happily, most of economic life is not like this.

Virtuous cycles

The quest for business success is the quest for a virtuous cycle. This happens when a player differentiates his product or service so that he can enjoy a higher margin than competitors, and yet have a larger market share than them; or when he achieves the same effect through having much lower costs than competitors (and therefore higher margins despite lower prices) as a result of superior scale, technology, cunning, or defensibly lower input costs. The player with higher margins can make further investments to consolidate and increase his lead; he can pay more to get the very best people or the most productive systems; he can afford to advertise or market at lower cost and higher efficacy; he can provide even better value and make the gap between himself and competitors almost unbridgeable.

This is what happens with successful, very profitable companies. This, you will recall, is the world of increasing returns. Firms with virtuous cycles always account for the majority of profits in a sophisticated economy.

Vicious cycles

Vicious cycles are the flipside of virtuous cycles, as seen from the viewpoint of the unsuccessful challengers. Those who are behind fall further behind. Returns diminish. St. Paul was a great exponent of virtuous and vicious cycles: he was forever writing that what goes around, comes around, for example 'Whoever sows sparingly will also reap sparingly; and whoever sows generously will also reap generously.'[5] Successful firms can afford to sow generously; less successful ones have to be more sparing.

When virtuous switches to vicious

The danger for successful firms is when something happens in the system to turn a virtuous cycle into a vicious cycle. How can this happen? One way is via the paradox of enrichment—you can have too much of a good thing.

Too much success makes you arrogant, complacent, or greedy. You ignore a new technology that has the potential to provide a better or cheaper service, because your success is build on the old technology. You make such fat profits that your managers or your unions raise the firm's costs beyond those of rivals. You stop listening to customers; you already know what they want, and they're always whinging anyway. You stop hiring new talent, or else you hire talented people, but stop them doing anything new. You use highly rated paper to take over other firms, paying more than the target is worth, and then you destroy value there by making them do things your way. You diversify into new products and services where you have no competitive advantage. You make the firm bigger, more complex, more heterogeneous, less manageable and less like a clan. If you do any of these things, nothing fails like success. This is the paradox of enrichment.

When the paradox of enrichment operates in business, there is a moral flaw. But with the second of our three power laws of caution, the law of entropy, there is an amoral physical force poised to derail our momentum.

The law of entropy

'Entropy' was coined by the German physicist Rudolf Julius Emanuel Clausius to mean the tendency of things to run down and wear out. He wrote in 1865:

> I propose to call the magnitude S [energy unavailable for work] the entropy of the body, from the Greek word [trope], transformation ... The energy of the universe is constant—the entropy of the universe tends towards a maximum.

People grow old. Houses fall down. Stars burn out. Cliffs slide into the sea.

The law of entropy is a restatement of the first two law of thermodynamics, developed in the years before 1850 in the quest to build better steam engines. French physicist Nicholas-Leonard Sadi Carnot found that when heat is lost, it is possible to get work out of the process. An Englishman, James Prescott Joule, discovered the converse: that when there is work, extra heat also arises. The first law of thermodynamics, arrived at separately by Joule and German scientist Julius Robert von Mayer, states that energy can neither be created nor destroyed—it can only change form. Then in 1850 Clausius, building on Carnot's work, gave us the second law of thermodynamics: any chemical system, be it solid, liquid or gas, will tend toward maximum disorder. Energy flows in one direction only, towards thermal equilibrium. Heat is transferred from one body to another, and this transfer cannot be reversed. Heat can only be used up once—it flows into the cooler body and can't be retrieved from it (without adding yet more energy). As the great Scottish scientist James Clerk Maxwell (1831–79) remarked,

> If you throw a tumbler full of water into the sea, you cannot get the same tumbler of water out again.

The law of entropy has parallels to two biological concepts that we examined in Part One: the red queen effect and the evolutionary arms race. The world changes, and to preserve what we had before we have to do more than we did yesterday. Things can be maintained, or even improved (paint from a house may fade, but it can be painted again better than it ever was), but it requires new action. A system's energy is discharged and lost to it, so life requires infusions of new energy.

To maintain success requires constant effort. The natural condition is not equilibrium: it's entropy. A company's competitive position rests on a bundle of unique resources and relationships that are alive and restive; like all systems and all relationships, if they are not tended, reinforced, and renewed, they'll falter and fall apart. It is entirely possible to counter entropy—how else could we have accumulated wealth in the remarkable way we have in the past 250 years?—but it requires constant innovation and improved use of the energy that is available.

Murphy's laws

Closely related to entropy are the laws attributed to 'Murphy.' They have no scientific validity, except perhaps as examples of entropy, but they certainly have resonance. They're useful for any successful organization, helping to puncture complacency and prepare for contingencies.

Murphy may have been apocryphal, like the ubiquitous Kilroy during the Second World War; or possibly he was Captain Ed Murphy of the Edwards Air Force Base, who said of an incompetent technician, 'If there's any way to do it wrong, he will.'

There is now a very large number of **Murphy's laws**. Here's a useful selection:

- If anything can go wrong, it will.
- If several things can go wrong, the one that will cause the most damage will go wrong first.
- If anything just cannot go wrong, it will anyway (for example, the *Titanic*).
- If you realize that there are four ways in which something could go wrong, and circumvent them, then a fifth way will promptly develop.
- Left to themselves, things go from bad to worse.

- ◆ If everything is going well, you have overlooked something.
- ◆ Nature always sides with the hidden flaw.
- ◆ Nothing is as simple as it seems.
- ◆ Everything takes much longer than you expect.
- ◆ It's impossible to make anything foolproof; fools are so ingenious.
- ◆ If the experts have spent a huge amount of time and failed to find the answer, it will be immediately obvious to the first unqualified person asked.
- ◆ When things go wrong somewhere, they go wrong everywhere.
- ◆ Whatever you want to do, you have to do something else first.
- ◆ Figures that are obviously correct will contain errors. A decimal will always be misplaced. The error will cause most damage to the calculation.
- ◆ If you get the premise right, but the argument wrong, you'll get the wrong answer; while if you get the premise wrong, but the argument right, you'll also arrive at the wrong answer. You are unlikely to get both the premise and the argument right.
- ◆ The probability of anything happening is proportional to the damage it will cause.

The law of unintended consequences

The third common way in which success turns to failure is through the unintended consequences of well-intentioned actions. Here there is simple miscalculation, based on failure to understand how systems operate.

Dietrich Dörner, professor of psychology at Germany's University of Bamberg, has written a fascinating book, *The Logic of Failure*,[6] which explores why intelligent people and institutions can proceed with care and goodwill and yet often produce disastrous results. He says that the problem lies in our patterns of thought, which are linear and take one thing at a time, thinking in terms of cause and effect. Because we don't think in terms of systems and their inter-relationships, we miss the big picture, pile small error on small error, and end up with spectacularly unintended and often tragic consequences. Dörner's ideas derive from systems thinking and clearly relate to ideas that we explored in relation to quantum mechanics (Chapter 6) and chaos and complexity (Chapter 9).

Dörner gives many examples of disasters. Why did the well-qualified engineers who planned the Aswan Dam and whose simple aim was to bring cheap electricity to Egypt not realize that the annual floods that they would stop had kept the Nile Valley rich and fertile for millennia? Why do planners of health programs in poor countries not take into account that increasing the numbers alive will also increase demand for food, and that without extra food production, improved healthcare will just lead to malnutrition and sometimes famine? Why did the operators of Reactor 4 of the atomic energy plant at Chernobyl, who had just won a safety award, end up with the ghastly explosion of April 26, 1986?

On a less horrific scale, what about the mayor and city council who dealt with traffic congestion and air pollution in a city by installing speed bumps and a 20 mile an hour speed limit? What could go wrong here?

Quite a lot. The cars had to travel in second gear, so they were noisier and produced more exhaust. Shopping trips took longer and the number of cars in the city center actually increased. But that was self-correcting. After a while, fewer people shopped downtown, preferring the big new mall on the edge of a neighboring town. That solved the noise and pollution problems, but led to many shops in the city closing. Tax revenues plummeted. Taxes had to be raised on the remaining businesses, which just reinforced the cycle of decline. All caused by a few speed bumps in a noble cause.

The theory of the second best

The example of the speed bumps links nicely to a theory beloved of economists and especially those in the area of public policy. This is the **Theory of the second best**, which says that reaching an optimal outcome in individual markets may lead to a suboptimal overall outcome. For example, if free markets led to an optimal position in all individual product markets, but left an economy with 40 percent unemployment, this would not really be optimal. The theory therefore says that instead of seeking optimality in each part of the economy, we should go for the best overall solution, which may imply 'second best' solutions in individual markets.

Stripped of economists' usual obsession with equilibrium and optimality, two very elusive goals, the theory of the second best is really just saying that the economy is a system, and that actions in one area may have unintended consequences in another. It is a useful idea because it tells us that we may have to compromise and that pursuit of one objective may be myopic: speed bumps must not be thought of purely in terms of their own objectives.

Incidentally, although the law of unintended consequences focuses on unfortunate outcomes, there are clearly positive unintended consequences, such as Adam Smith's 'invisible hand':

> *Every individual necessarily labours to render the annual revenue of the society as great as he can. He generally, indeed, neither intends to promote the public interest, nor knows how much he is promoting it … By pursuing his own interest he frequently promotes that of the society more effectively than when he really intends to promote it.*[7]

System dynamics

Jay Forrester of MIT was a computer pioneer who developed **System dynamics** in the 1960s and 1970s (known by other practitioners as systems thinking and developed since about 1950). Forrester was one of the first to call attention to the unintended consequences of well-intentioned policies on issues such as urban decay or the environment. Typically, he said, the policies attacked the symptoms of the problems, alleviating the symptoms but often exacerbating the fundamental problems that were 'systems' rather than discrete issues.

Systems thinking has much in common with the concepts of chaos and complexity. The intention in all cases is to identify the underlying system that operates, in order to find long-term solutions rather than short-term palliatives.

Avoiding unintended consequences

Dietrich Dörner suggests the following prescriptions:

◆ *Set clear, explicit, positive, and multiple goals.* If possible, we should for-
mulate concrete goals. (If we can't, muddling through is better than
inaction.)

◆ *Pursue several goals at once.* If you focus on one goal alone, you will produce all kinds of unintended by-products. You may object that to pursue several goals at once may bring conflicts between the goals. This
is true. But the conflict is constructive, because it forces us to consider
the relative priorities and trade-offs implicit in the goals.

For example, would we rather have speed bumps and a clean,
quiet, pollution-free city center, or a thriving retail and business center at the expense of some congestion and exhaust fumes? Can we
think of any realistic way of having most of our cake, and also eating
most of it?

We can't always realize all our goals at once, because the goals
may partly conflict with each other. We must be prepared to compromise. We should always have a clear set of priorities, but be willing to
change them if it's clear that they'll lead to results we don't really
want.

◆ *Construct hypotheses and test them.* If we do *x*, it will result in *a*, *b*, and
c. If we like *a*, *b*, and *c*, we may try out *x*. If it doesn't have the expected
result, at least we have more data. Wrong hypotheses can be corrected.

◆ *Use analogies to go from what you know to what you don't.*

◆ *Think of everything as a system* and try to identify all the important system elements. Form a model of the system. You can start with a single element but then think of its context. In the pool example, you
could start with the fish. A fish breathes, it eats, it excretes. It needs
oxygen. What happens to the waste? We can begin to see how everything in the system fits together.

◆ *Think about problems you don't have at the moment but which may emerge
as side-effects of your actions.* Think about what may happen over time.
Imagine potential pitfalls.

◆ *Don't hastily ascribe everything that happens to one central cause.* This is rarely the case.
◆ *Construct simulations.* By playing games, with many variables affecting a system, you'll learn how systems work, and be able to make mistakes with no real-life penalties.

One thing that Dörner does not say but that seems apparent to me is that, ultimately, the most effective antidote to unintended consequences is human creativity and adaptability. Unintended consequences arise because we live in non-linear systems and because we make changes, because we *do* things. We are restless, just like nature, just like evolution. Each action generates new instabilities, and will always do so. Unintended consequences can never be eliminated. We should be aware that they will arise. We should be ready to notice them before they have done too much harm. And we should be creative in correcting them (and aware that the corrections will lead to further unintended consequences, which will require further correction…).

How to perpetuate success

Dietrich Dörner's model and examples are useful for those in charge of a successful business; they help us to think about what could go wrong. Consider this in three ways:

◆ What happens if something in the business system changes unilaterally?
◆ What new actions that we've taken in the current business system could have had unintended consequences?
◆ How do we plan new initiatives that will be as successful as current ones and that won't have unintended consequences?

What happens if something in the business system changes unilaterally?

A successful system can only turn into an unsuccessful system (from our point of view) if one of two things happens: either we do something differently, or something else in the system changes to our disadvantage.

Therefore, if we're doing the same as ever, and things start to go wrong, it must be because of a change (or changes) in other elements of the business system.

So start by asking: What has changed? Have the customers changed what they want? Is a competitor gaining market share? Why? Is the technology or the business definition shifting? Has everyone else found a way of cutting their costs while ours remain as they were?

Construct several hypotheses, both complementary and competing. Remember that there's unlikely to be one simple cause. Even if there is, it will have second- and third-order effects that must be traced.

Test and refine the hypotheses, until there's a reasonable chance they're right. Then act to restore your advantage. If this doesn't work, go through the whole cycle again.

What new actions that we've taken in the current business system could have had unintended consequences?

What are we doing differently? If you know the answer, fine. If not, ask other people. Introspection will not be accurate or complete.

Map out all elements of the business system, including (but not necessarily confined to) your customers (including any important different types of customer), suppliers, distributors, other collaborators, colleagues, cost structures, technologies, regulators, and so on. Imagine all possible impacts that the changes may have had on each part of the system, and the consequential results, particularly bad results, that might have arisen.

Construct and test hypotheses. Be suspicious of pat solutions involving just one variable. For the system to have changed fundamentally, several aspects of the system are likely to have shifted.

Then act, and if it doesn't restore the system to your advantage, go through the cycle again.

How do we plan new initiatives that will be as successful as current ones and that won't have unintended consequences?

Here we need to interject skepticism from outside the Dörner model. The chances are that new initiatives won't be as successful as existing ones. The only reasonable basis for thinking that they might be as successful is if they use the same formulae, skills, competencies, technologies, and any other key attribute (such as a fantastic proprietary client base) that drives the success of the existing business.

But let us assume this is true. What then? Well, the Dörner model is good for thinking about where there might be unintended consequences. In particular:

◆ What unintended consequences could the new business have on the existing one? Think through all elements of the system and their relationships to each other.
◆ What other unintended consequences could entry to the new business have? Again, trace through all components of the new business system.
◆ If there do turn out to be negative consequences, construct hypotheses and test them (as discussed above) until you have an answer that works.

Summary

Sustainable success is built on a virtuous cycle. But success often contains the seeds of its own destruction. A virtuous cycle can turn into a vicious cycle if success changes the conditions that led to success in the first place. This is the paradox of enrichment: riches corrode the will to please customers. Successful firms become flabby and complex. Enter greed, complacency, or arrogance; exit success.

Even without such an ethical failure, two other dangers lurk: entropy and unintended consequences. Like everything else, the elements of business—ideas, technologies, individuals, teams, and corporations—are subject to entropy. Without maintenance and renewal, all these elements will wind down. Business conditions are always becoming more stringent. Everything is always becoming something else. Entropy is hardest on yesterday's winners. Success has to be reinvented, yet past success makes future success difficult. Firms that have extracted more value than they have added will be especially vulnerable; their pool of collaborators will dry up.

Unintended consequences may also intervene. Business is a system of inter-related components. Change in any one component may go unnoticed, yet break the spell of success. We need to anticipate unexpected consequences. When they arrive anyway, we must model their causes carefully.

Ultimately, sustaining success is a moral issue. Although we might wish it were otherwise, business is the art of pleasing the customers we choose, with more flair, economy, and style than anyone else. This is a hard calling. Those who are successful have great *technical* advantages, but equally great *attitudinal* disadvantages. Can they display the greatest flair, economy, and style, when conditions and competitors change? Can success sustain superior service? Can success sustain simplicity? Can success keep focus? Can success keep lean?

Experience replies: Sometimes. Not often. Not for ever. Not without the perpetual reinvention and delivery of superior value.

Action implications

◆ *Sustain success by creating new value every day.*

◆ *Keep yourself and your firm humble, service oriented, focused, lean and hungry.* Eschew corporate complexity. Root out arrogance, greed, and complacency.

◆ *Expect and correct unexpected consequences.* Learn to anticipate and deal with them. Think of business as a system whose components are always shifting, where a major shift in just one component can change the entire system, where there will always be unintended consequences, and where continual monitoring, adjustment, and creativity must be deployed to detect and overcome them.

Notes

1 Richard Pascale (1990) *Managing on the Edge*, Simon & Schuster, New York.

2 Arie de Geus, *The Living Company*.

3 Ellen de Rooij (1996) 'A brief desk research study into the average life expectancy of companies in a number of countries,' Stratix Consulting Group, Amsterdam, quoted in Arie de Geus, *The Living Company*.

4 See Karl Sigmund (1993) *Games of Chance*, Oxford University

Press/Penguin, Oxford/London, p. 45.

5   *Second Letter of St. Paul to the Corinthians*, Chapter 9, verse 6.

6   Dietrich Dörner, *The Logic of Failure.*

7   Adam Smith (1776) *An Inquiry into the Nature and Causes of the Wealth of Nations*, Book IV, Chapter 2.

Part Three Concluding Note

One common theme of Part Three, shared with Parts One and Two, is the wonkiness of the world. We expect and search out linear relationships and rejoice when we find them; but we tend to ignore the more frequent non-linear relationships, because they are inconvenient and perplexing. Yet the message of Part Three is that non-linear relationships can be understood and be extremely useful.

The concept of chaos is very helpful because it tells us the importance of 'initial conditions': markets, relationships, and corporations evolve as they do because of early, chance events, and quickly get frozen into the patterns formed at the outset. Chaos also tells us about the fractal nature of business, which is an insight that, properly understood, can save a great deal of trouble.

Complexity theory demonstrates how systems emerge and organize themselves into something different from their component parts: a phenomenon at once awesome, constructive, and destructive.

The 80/20 principle is also a fantastically useful power law, showing that a particular sort of wonkiness is deeply ingrained in the universe and enabling us, almost infallibly, to extract more from less.

The non-linear, jerky nature of market growth and technological change is also something that, when appreciated, can help to separate the fad from the trend, tell us whether we can drive a new product or business across its tipping point, and enable us to spot what may become a mega-success. Insight into the nature of networks and electronic business can also tell us when there is scope to create enormous new value, and how to be on the right side of the extremely asymmetrical capture of value created.

Finally, the paradox of enrichment, the law of entropy, and the law of unintended consequences all highlight the systemic pitfalls of success and how to sustain success by systematic avoidance of the pitfalls.

A second common theme of Part Three is the tension between *laissez-faire* and intervention. In non-linear systems we deal with some extremely powerful natural forces. We have seen many different kinds of 'invisible hands' at work, sometimes producing extremely pleasing results, and sometimes infuriatingly defeating our best-laid plans. As with the forces of biology and physics, however, we must strike a balance between the opposite errors of automatically accepting the power laws on the one hand, and ignoring them on the other. The laws are there, and we had better take note. But might is not right; what is, is not what should be; nature is not infallible or inherently virtuous, and neither are markets, high-growth phenomena, successful corporations, or self-organization. Progress requires us to channel nature and its forces, to use them for our own ends, and to intervene when they threaten civilization and its fruits.

Part Three's third and final theme is that business, and the forces operating within and around it, are not essentially different from the rest of 'life, the universe, and everything.' All the power laws apply to life generally; they apply to business because business is part of life. Hence, we can use the insights gathered beyond the walls of commerce. But it also follows that the view of business as a separate terrain—an enclave with its own conventions and laws, a mysterious no-go area for the rest of society, a quasi-medieval guild with its own governance, a field of study that requires its own schools, a landscape that can safely ignore more general insights on how to live a happy and fulfilled life and behave responsibly to others—is deeply flawed. Business is an intrinsic part of the messy reality governed by non-linear forces.

We have had the boundaryless corporation and the boundaryless market. Now it is time for boundaryless business, boundaryless knowledge, boundaryless technology; for business and science and knowledge naked before humanity and society; for a view of how things work, and how we can be successful, that applies equally to business and all other parts of life.

Part Four

So What?

Introduction to Part Four

The Finale answers the question: "So what?" by asking where the science of the past four centuries has taken us. In one sense, we are left with more questions than answers: the more we know, the more we realize how little we know.

Finale

On the Gospel According to the Power Laws

We are better at predicting events at the edge of the galaxy or inside the nucleus of an atom than whether it will rain on Aunty's garden party three Sundays from now, because the problem turns out to be different ... It is the best possible time to be alive, when almost everything you thought you knew is wrong.

Tom Stoppard, *Arcadia*

All aspects of business—all products, all activities, all methods—have an information structure at their core that has long been hidden, just like the genetic code of plants ... executives will have to create new genetic structures for their businesses.

Jay Walker, internet entrepreneur

The scientific laws driving progress

In 1859, Charles Darwin closed his account of evolution by natural selection with a skillful contrast between the exuberance of nature, and the economy of the scientific laws that had produced such a pleasing result:

It is interesting to contemplate an entangled bank, clothed with many plants of many kinds, with birds singing on the bushes, with many insects flitting about, and with worms crawling through the damp earth, and to reflect that these elaborately constructed forms, so different from each other, and dependent on each other in so complex a manner, have all been produced by laws acting around us.[1]

Some things we do know

Darwin's account of evolution by natural selection is a marvel of inference and insight, where God is gently shunted to the sidelines, and creation from the first form of life to the rich complexity of countless species is reduced to one simple, dialectical process: growth via sex, inheritance, variation, a ratio of population increase so high as to lead to a struggle for life, and therefore to natural selection, divergence of character, and the extinction of less-improved forms.

In reviewing some of the most important triumphs of science from the last four centuries, we've been able to see yet more wonderful things than those revealed by Darwin, but equally capable of reduction to a few simple power laws.

We've looked at the remarkable triumphs of physics, which have shown not only how the same rules of motion apply on heaven and earth, but also the astonishing way that the smallest parts of matter operate, and the awesome power that can be generated by such understanding. We've seen that space and time are not two separate dimensions, but are intimately linked.

We know how Darwin's evolution by natural selection operates, and the incredible information power packed in exactly the same way into the genes of all living organisms. We know that genes replicate themselves by using organisms as vehicles. We suspect that our genes are not fully aligned either to our own objectives or to commercial, urban society. We see human appropriation and creation of knowledge as an alternative and additional form of replication to that of genes, one that may enable humans to drive parts of the evolutionary process in the direction we want.

We have some insight into how complex systems can emerge from simple ones, how everything in the universe tends to organize itself into systems, and how even the most puzzling phenomena follow intricate and predictable patterns, at once similar to and different from each other. We know that in any distribution of a population, whether of people, clouds, diseases, events, or anything dead or alive, whether good or bad or neutral from our viewpoint, a small minority of the forces will have much more influence than the great majority, and we know how to distinguish the vital few forces from the trivial many.

Mathematics has given us dazzling insights into the power of exponential growth and how the same patterns recur in numbers, regardless of the phenomena being observed.

We know how new technology evolves in a lumpy fashion, and how to predict when we're likely to move from one dominant technology to another. The same tools enable us to observe, and often predict, how and when any phase transition will take place, whether we are talking about major social change, a trend or fashion, an epidemic reaching crisis proportions, or a company's profits taking off.

Although economics and the social sciences have generally been a disappointment, the last three centuries have taught us a few very useful things: how markets and organizations are self-organizing, dynamic systems, with their own way of sorting things out; how wealth is created by the division of labor and trade, based on comparative rather than absolute advantage; how returns can increase over time, so that costs and prices can go down for ever; how networks increase value; how economic arbitrage creates more from less, as resources are shifted from low- to high-productivity uses; and how societies evolve by means of increased specialization, reciprocity, trade, technology, and ever greater levels of cooperation and interdependence.

Some things we know we don't know

We also know more about the limits to our knowledge and perception. We know what it is impossible to know for sure, a disconcertingly wide domain. We know that there is no objective truth, and that we are

continually distorting and adding to reality. We see that the most important things we can create are concepts, ideas, and hypotheses, and that
these take on a life of their own: information and imagination are our
evolutionary aces in our game with the inscrutable universe, which hith-
erto has monopolized the best cards.

We know that uncertainty is at the heart of the universe. Although we
can find laws describing what happens, they are stochastic rather than
deterministic: we need to think in terms of probabilities rather than cer-
tainties. We know that chance is central to all life.

We know that we humans sit in a very odd relationship to our environment: we suspect that our genes are out of sync with the society we
have created, and that our emotions have not caught up with our reason.
We are not sure whether we are controlling our genes, or they are controlling us: the chances are that an intriguing battle is on, both for each
individual and for society as a whole.

We also don't know whether our universe is the only one with 'intelligent life,' whether there are other universes from which we may have
branched out in space and/or in time, or how much longer earth and the
universe will last.

Oh, and yes, Tom Stoppard is right: we don't know whether it will rain
at Aunty's garden party.

Darwin would have been amazed at how much more we know, and
at how much less.

The power laws change our perspective

The most interesting modern science is that which unifies its different
branches, enabling us to glimpse universal power laws. But there is a more
fundamental unification that is beginning to take place. As the great
Harvard biologist Edward O Wilson says:

*The greatest enterprise of the mind has always been and will always be the
attempted linkage of the sciences and the humanities.*[2]

Wilson also argues for 'a belief in the unity of the sciences—a conviction, far deeper than a mere working proposition, that the world is orderly and can be explained by a small number of natural laws.' Through study of our power laws, I think we can see a few themes that link together the sciences, the humanities, and business.

The power laws cast business is a new light. Without being fanciful, we can see that business operates in the same way as other complex systems, and is subject to the operation of laws as other parts of the universe. We have described the most important of these laws, and there are not too many of them: evolution by natural selection, genetic laws, Gause's laws, evolutionary psychology, the prisoner's dilemma, Newton's laws, relativity, quantum mechanics, chaos, complexity, the 80/20 principle, punctuated equilibrium, the tipping point, the law of increasing returns, the paradox of enrichment, the law of entropy, and the law of unintended consequences. Of these, only the prisoner's dilemma and the law of increasing returns apply *specifically* to business and other human affairs; for the rest, the laws apply to business because they refer to nature and life and business is part of life. Of the 17 major power laws, there are six that are even more important than the rest, and in five of the six cases it is clear that the laws apply to business in precisely the same way as they do to the rest of the universe; in the sixth case, genetics, the laws apply in a similar but slightly different way.

These six pre-eminent power laws are: evolution by natural selection, genetic laws, Gause's laws (these three forming a natural cluster of biological laws that fit perfectly with each other); and chaos, complexity, and the 80/20 principle, which comprise another snugly fitting cluster of non-linear laws.

The biological laws

Business evolves by selection in the same way that nature does: both progress through variation, innovation, the rejection of most new variants, and the propagation of a few very successful variants. Generalities generate differentiations. Differentiations become generalities from which other differentiations emerge. Development depends on co-development—in nature, of other genes and organisms; in business, of other elements of information, and of business vehicles, animate and inanimate. The odds against survival are high, leading to a struggle for life. The conditions of life determine whether species and individuals survive. Finally, the process of natural selection is arbitrary and random as well as effective.

The best guarantee of survival and good fortune is to have good genes. Genes are powerful packets of information that seek to replicate themselves as widely as possible by finding appropriate vehicles. These vehicles coat the genes—which are essentially massive forces of information, packed into a tiny and fragile physical shell—with robust physical protection. Good genes gravitate to the best vehicles for them and their fellow genes.

Business has a parallel genetic process. Business is driven by ideas and information ('business genes'), which seek vehicles for their expression and replication. These vehicles include people, technologies, products, services, corporations, and all the physical paraphernalia of the modern economy. Everything written in the previous paragraph about genes applies precisely to business genes.

The difference between biological genetics and business genetics lies here. Organisms are passive recipients of genes. In business, there is a degree of choice available for proactive individuals and groups of people: they can identify the best business genes, and even create new business genes. Yet this difference is not really one between business and nature; rather, the difference arises because humans are an odd part of nature, one that can create 'memes'—essentially humanity's own version of genes, ideas that can be replicated in the same way as genes. As humans are part of nature, and business is an extension of humans, business genes, like business itself, can be viewed as part of the seamless web of nature and the universe.

Gause's laws highlight specific lessons of evolution and genetics. With limited resources, a species or individual must be different from others to survive. Differentiation allows us to find different ways of making our living, and therefore avoid direct confrontation for the same food and space.

The non-linear laws

Important parts of life and business are 'non-linear systems,' where cause and effect are tangled together; patterns exist but they are irregular and often far-reaching. Chaos tells us the importance of 'initial conditions' in determining outcomes, and therefore in grabbing the first-mover advantage. Chaos also tells us that business (like clouds, trees, or coastlines) is 'fractal'—each type of business has its recurrent patterns and rules for success; recognizing them requires respect for the differences, experience, skill in pattern detection, and specialization.

Complexity reveals how complex systems such as amoebae, Silicon Valley, or meteorites emerge automatically, as if by an invisible hand. Because business is a complex system, it is always dynamic and always adjusting itself to new realities. Chaos and complexity also demonstrate the role of chance in business, and therefore the importance of flexibility and dedicating some resources to long-odds strategies.

The most useful of the non-linear laws is the 80/20 principle. This shows how a few very powerful forces nearly always determine most of what happens. Therefore, smart or lucky creatures find the powerful forces and ignore the great bulk of existence that fills the universe with insignificance.

The power laws present a coherent view of reality

The biological laws and the non-linear laws complement each other and are also consistent with each other. Evolution is an extremely non-linear process; it is an example of chaos theory in action. Species (and combinations of species) emerge as complex systems. Evolution is also the best and most important example of the 80/20 principle.

The other power laws also cohere with the biological laws, the non-linear laws, and each other. For example, relativity and quantum mechanics have many parallels with chaos and complexity; evolutionary psychology flows from theories of evolution and genetics; the prisoner's dilemma, and associated theories demonstrating the importance of cooperation, resonate strongly with evolution and other non-linear systems;

punctuated equilibrium is an evolutionary theory that has a clear parallel in the 'economic' theories of growth via technological change and in the tipping point; the tipping point and the theory of increasing returns relate closely to the 80/20 principle; and the law of unintended consequences can be viewed as a corollary of chaos theory.

The knowledge embodied in the power laws is itself 'on the edge of chaos,' poised between coherent theories with supporting data marshaled in good order on the one hand, and open-ended speculation with many loose ends on the other. If theories are to be really powerful, how could it be otherwise? Some order is necessary for a law to be useful; but some disorder is also necessary, to allow us to improve the state of our knowledge and to reach out to other realities, both known and currently unknown.

What most impresses me about the power laws is the consistency between their *gestalt* and value in a non-business context, and their application to business itself. The power laws help us understand 'life, the universe, and everything,' but they also simultaneously help us understand business, and show how business is not so very different from other aspects of life. Business is not an alien planet; it is part of the texture of life on earth, and the rules for success are very similar to those in the rest of existence.

So what are these rules for success? How do the power laws change our view of business?

Business is driven by ideas and information

◆ Out goes the view that 'management' and 'business corporations' are at the center of value creation in business.

◆ *In comes the view that we should look further back. Value is created by powerful business ideas and technologies and we are just swept up in the backwash from these. Managers and corporations are mere chess pieces—and usually just pawns—in the flow of information into acts of physical creation.*

◆ Out goes the view that the view that corporations create value and profits by establishing competitive advantage.

◆ *In comes the view that information drives value, profits, and competitive advantage; corporations that appear to have competitive advantage just have temporary and non-proprietary access to better information, which drives value and profits.*

◆ Out goes the central role of corporate competition.

◆ *In comes the view that the fundamental unit of value in business is economic information, comprising business genes. The driving forces are science, technology in the broadest sense, other forms of useful information, open markets, and people, as individual scientists, technologists, entrepreneurs, and executives. Every business is an information business and succeeds to the extent that it is a superior vehicle for powerful information.*

◆ Out goes the view that business progress happens through corporate competition, and that this process resembles evolution by natural selection.

◆ *In comes the view that, except for very small and young businesses, corporate competition is a phony war. In general, the struggle for life bypasses corporations, or takes place unseen within them. The important struggle for life occurs among business genes and simple self-organizing systems; among competing technologies and subtechnologies, products and product components, and in the way that products and services are conceived, constructed, and delivered.*

Value is created where the struggle for life is greatest. But value is also captured, by corporations and people, where the struggle for life is least.

The value creation process proceeds via innovation and natural selection

◆ Out goes the view that innovation is a minority activity.

◆ *In comes the view that innovation is the essence of business and that innovation proceeds via constant variation, selectivity, specialization, and experimentation. If you aren't experimenting, varying, and changing things constantly, then you aren't creating value and you won't be successful.*

◆ Out goes the view that corporations can control their own destiny and determine their own success.

◆ *In comes the view that individuals and corporations are part of a random process where change in the only constant; where influence on events is possible but control is not; and where success, even when deliberately engineered, is as much an accident of the moment as the result of foresight, cunning, and skill. Success is never an annuity in perpetuity. Success is always achieved by alignment with greater external forces. We may think that we are harnessing such forces, but more likely we are being used by them.*

◆ Out goes the view that most experiments should succeed.

◆ *In comes the view that most experiments should fail.*

A new information medium threatens to bring about punctuated disequilibrium

◆ Out goes the view that business proceeds according to immutable laws of economics and strategy.

◆ *In comes the view that economic and strategic laws last a very long time, but that occasionally the business context changes so dramatically that the laws need to be reframed for the new context. The nature of business changes fundamentally when there is a massive shift in either a dominant communications medium (e.g. writing, mass printing, television) or a dominant technology (e.g. agriculture, the steam engine, the assembly line, information technology). Probably, the internet represents both a new dominant technology and a new dominant communications mechanism. If so, it will be the biggest disruption to business equilibrium since the steam engine.*

◆ Out goes the view that business has more power than consumers, and that big business has more power than small business.

◆ *In comes the view that the consumer and the entrepreneur will gain power at the expense of big business.*

Variation creates uniqueness; uniqueness creates monopoly

◆ Out goes the view that you should gain a lead over your competitors.

◆ *In comes the view that you should create your own new business space, where there are no competitors and where the gravity of competition cannot depress returns.*

◆ Out goes the quest for monopoly by dominating accepted business models.

◆ *In comes the quest for monopoly by creating uniqueness.*

Finding the right *growth requires great skill and selectivity*

◆ Out goes the view that it is difficult to find growth.
◆ *In comes the view that most growth is futile and profitless.*
◆ Out goes the view that growth comes from entering new markets.
◆ *In comes the view that profitable growth comes from creating new markets out of promising raw material that is lying around unused. Profitable growth arises by identifying technologies, ways of doing things, and social trends that are accelerating fast, yet have not quite reached their tipping points.*
◆ Out goes the view that growth is a matter of mobilizing corporate resources to conquer new worlds.
◆ *In comes the view that growth comes from creating new value for customers. New value is created from better ideas. Better ideas use fewer or cheaper resources to satisfy certain customers more.*
◆ Out goes the view that vertical integration is sensible, or that corporations should operate in most or all elements of the value chain.
◆ *In comes the view that there are sweet spots in every industry, which give the majority of the profits of a value chain through participation in only a small number of its activities.*

Less is better

◆ Out goes the view that more is better.
◆ *In comes the view that less is better (and that, very often, less is more).*
◆ Out goes the view that most effort is (or should be) rewarded.
◆ *In comes the view that most effort is wasted. Most big results come from small proportions of input.*

Don't build cathedrals; do build open networks with increasing returns

◆ Out goes the view that corporations should own strategic assets and that high profits require extensive control.
◆ *In comes the view that corporations should own and control less, and influence more.*

◆ Out goes the view that potential acquisitions should be high on the management agenda.

◆ *In comes the view that formal or informal alliances between corporations cost much less than acquisitions and yet can deliver almost as much benefit, and sometimes more benefit, than acquisitions.*

◆ Out goes the law of diminishing returns.

◆ *In comes the law of increasing returns. The best example of increasing returns is information itself: good information always has increasing returns and infinite marginal returns. Information does not wear out or degrade when it is used; instead, it increases in reach, depth, and value. Information once sold can be resold indefinitely. If the number of users of information increases for ever, its cost decreases for ever.*

◆ Out goes the view that corporations and markets should resemble cathedrals, or forts.

◆ *In comes the view of markets as bazaars open to all comers, and corporations as traders in bazaars.*

◆ Out goes the view that corporate secrets must be protected.

◆ *In comes the view that secrets not exposed to use beyond the corporation will soon cease to be useful.*

◆ Out goes the view that corporations should avoid leakage of their knowhow.

◆ *In comes the view that some leakage is* desirable, *since leakage in requires leakage out, and leakage in is invaluable.*

◆ Out goes the view that corporations should create and defend their own proprietary business systems.

◆ *In comes the view that corporations should cooperate with other corporations and users to create common, open networks.*

Use both *reason* and *passion*

◆ Out goes the view that business is largely a matter of rational calculation. Out goes the view that the business world is largely composed of linear, cause-and-effect relationships.

◆ *In comes that the view that business is* both *a matter of rational calculation and a matter of irrational passion. In comes the view that creativity is both a rational* and *an irrational process. In comes the view that business context is* both *linear* and *non-linear, in roughly equal proportions.*

◆ Out goes the view that there is always one dominant route to success.
◆ *In comes the view that there are always multiple routes to success.*
◆ Out goes the view that business is always black, white, or grey.
◆ *In comes the view that business is always colorful, and that it can be* both *black* and *white at the same time. The opposite of a great business truth is not a fallacy. The opposite of a great business truth is another great business truth.*

Back the favorite heavily and a few outsiders lightly

◆ Out goes the view that business is a money-making machine.
◆ *In comes the view that business is a book of bets.*
◆ Out goes the view that a firm should place several large bets. Out also goes the view that you should bet the company on one proposition.
◆ *In comes the view that a firm should have one main bet, but also a number of small bets at long odds.*

Sustain success by gaining and giving loyalty—and by continual improvement

◆ Out goes the view that nothing succeeds like success.
◆ *In comes the view that nothing fails like success: that enrichment, entropy, and unintended consequences are endemic, and that each day brings a fresh struggle to create new value and hence perpetuate success.*
◆ Out goes the view that competition in a series of discrete transactions is at the heart of a market economy, and that learning how to compete effectively is one of the most important things for corporations and executives.
◆ *In comes the view that competition is vital for an economy, but that competition largely takes care of itself; that business is a series of related transactions, linked together by cooperation, loyalty, networks, serial reciprocity, and reputations; and that learning how to cooperate with the best cooperators is the single most important individual and corporate competence.*

The gospel according to the power laws

In the beginning was information. Each day brought, and brings, more and better information. All business is information—the gathering, creation, refinement, combination, processing, and delivery of information. Information goes into products and services. But the information is not consumed; rather, new information is created. Information is retained and enhanced, alive and bubbling, in the brains of businesspeople and in the networks and vehicles set up to provide goods and services.

The universe is restless, dynamic, ever changing, expanding. Information begets information—more information, better information; more diverse information, more specialized information, more accurate information. The universe is endlessly creative, endlessly destructive. It makes mistakes, corrects mistakes, and then corrects the corrections, which themselves contain mistakes, which require correction… in an endless cycle that always increases richness, but never reaches perfection. Information can never be complete, never be consistent, and never be absolutely true.

Business exists to satisfy and create human needs and wants, to create and enhance civilized conditions of life. Businesses thrive if they do this well and differently. But, happily, they will never do it perfectly. The business universe can therefore expand for ever, because there is always room for something more and something better.

All progress requires improvement: a new product or service, or the delivery of existing ones in cheaper, better, or more convenient forms. Improvement requires experimentation, variation, and market exposure.

Really successful businesses meet three conditions. They are different from all other businesses. They make better use of ideas and resources. They continually improve; they use myriad experiments to ensure that they remain different from any other business. You can't catch a moving target that is continually creating its own new space.

Most experiments fail. We should let them. We should concentrate energy on the few successful experiments. We should conduct new experiments on these successful experiments, so that there are always new variants of them. For continued success, this process must never flag.

Business is exciting and challenging because new and better information is always available. New ideas, new ways of doing things, new potential partners, new customers, and new demands from existing customers all create a kaleidoscope of potential change and improvement.

Technological change drives growth. Technological improvement is not just inventions and the application of sophisticated science, but also the use of all kinds of knowledge to make things better and cheaper. Every successful businessperson is a technologist, using and creating knowledge, which others then use for further improvements.

Technological change can be spotted accelerating along the runway, before the take-off. Innovators need keen eyes and fast skates; but they do not need to start rich.

Change is blocked by three things—the failure to recognize, collect, and use information; the inbuilt human reluctance to take risks; and the tendency to build corporate fortresses that are larger, more diversified, and more isolated than they should be. All three blockages create great opportunities for entrepreneurs.

A narrow canvas is usually better than a broad one. But the more narrow the focus, the wider must be the window on developments elsewhere, and the network of weak ties. The ideal? Focus without high walls. Specialization without inflexibility. Differentiation without hubris. A unique stall in the bazaar, not a cathedral on the hill.

Business abounds with profitable asymmetry. A small minority of effort produces a large majority of value. Some things are *much* more profitable than others. It's much more valuable or economical to do things one way rather than another. It's much more productive to work with some individuals, some teams, and some networks than with others. The most productive resources are distinctive, and are committed to constant change and improvement.

These are the rules for business revealed by the power laws. They are your route to success. They show that, always, entrepreneurial bonanzas lurk unexploited. There are always new combinations of ideas, technologies, fellow travelers, suppliers, distributors, customers, and partners that you can use to create a superior business system. There is always a way of doing something better, and of finding something better to do.

Now all you have to do is to do it.

Notes

1 Charles Darwin, *On the Origin of Species by Means of Natural Selection*. I am quoting from the start of the final paragraph and from the 1985 Penguin, London version edited by J W Burrow, p. 459.

2 E O Wilson (1998) *Consilience: the Unity of Knowledge*, Alfred A Knopf, New York/Little, Brown and Company, London. Now available in an Abacus paperback (London).

Index

Acknowledgments

I have very much enjoyed writing this book, mainly because it has introduced me to a new world: that of the physical sciences. Although I had dabbled a little in biology, I had no idea how thrilling the encounter with the science of the seventeenth to twentieth centuries would be: both the ideas themselves, which tell us how the universe is governed, and the way in which they were discovered, causing consternation to scientists and often disbelief among the educated public. My first debt is therefore to all the scientists and writers whose insights I have appropriated, and especially to the sources quoted in the references.

My greatest debts to contemporaries are to Richard Dawkins, a brilliant scientist whose combination of Darwin and modern genetics is an unforgettable marvel, and who writes like an angel; and to Matt Ridley, a science writer who generates more intriguing juxtapositions and insights per page than anyone else I know. Their ideas have helped immeasurably in the development of my theory of business genes—a particular subspecies of the 'memes' invented by Dawkins—and I only hope that they do not think my elaboration of their ideas unworthy or trivial.

Very late in the writing of this book, I was given an advance copy of Jane Jacob's excellent short book, *The Nature of Economies* (Random House, New York, 2000), which I found highly congruent with my own argument, but also extremely useful for refining some of this book's themes.

I would also like to thank all my friends who have made useful comments on various large manuscripts, and especially Mark Allin, Dr Richard Burton, Robin Field, Anthony Hewat, Dr Peter Johnson, Clive Richardson, and Patrick Weaver, who have been extremely generous with their time and added many insights, and, at least as important, subtracted many of my 'insights' that were insufficiently grounded in the power laws. Dr. Marcus Alexander of the Ashridge Strategic Management Centre gave me a rigorous review and also access to his exciting research on 'boundaries.' In a class all of his own is Dr. Chris Eyles, a vet turned internet strategist, who has encouraged and badgered me throughout the process, adding his own blend of wisdom and knowledge.

Chris also supplied the structure for the book when this issue was driving me nuts—thanks, Chris, now get on with proving that the internet really can make money.

Two absolutely crucial partners in this enterprise have been my researcher, Andrej Machacek, of Balliol College, Oxford, and Nicholas Brealey, my publisher. Andrej did all the difficult initial research, telling me what to read, dredging up obscure monographs on some important subjects, and summarizing hundreds of power laws with amazing brevity and accuracy. He has also been a charming and valuable foil as we have debated the ideas in the book, as well as reviewing all six drafts and making wise comments, some of which I have incorporated into the text. Anyone who wants a terrific researcher could not do better than contact Andrej.

Nicholas Brealey is an anomaly: a publisher who actually cares passionately about the ideas he publishes, and who contributes original ideas as well as devastating comments on the text. It was also Nicholas who suggested the subject for the book, following the success of *The 80/20 Principle*, pointing out that it was a power law and that there were several other key laws of the universe. Nicholas is a phenomenon and it is wonderful working with him. Sally Lansdell and Sue Coll also made major contributions to structuring the book and making it easier to use. Eileen Fallon, of the Fallon Literary Agency in New York, provided a terrifically valuable critique of the penultimate draft of the book. Many thanks, Eileen.

I also am extremely grateful to my assistant, Aaron Calder, who has greatly speeded up the production process. I must admit that before Aaron came along I was using an antique version of Word Perfect in DOS, whereas under his tutelage I have become proficient in using Microsoft Word and Windows, truly one of the wonders of the twentieth century. Aaron has also performed many more recondite editorial functions that are still beyond my ken, as well as keeping me amused on the dark days when I wondered if a coherent book would ever emerge. (Perhaps it hasn't, but at least it got past Nicholas.)

Finally, my profound thanks to Lee Dempsey, my partner, for his daily love and support.